D1293001

WELLS HOUSE ROAD NW10

FALCON ROAD S.W.

MOSELLE STREET N.17

WAPPING PIERHEAD
Nos 4½ to 11.
P 6 34

OLD FLEET
LANE EC4
CITY OF LONDON
Formerly Fleet Lane

KNIGHTSBRIDGE SW
CITY OF WESTMINSTER

BROOK STREET N.17

MUSWELL ROAD N.1

SURREY CANAL
ROAD
London Borough of Lewisham SE14

WINTERBROOK ROAD SE24
LONDON BOROUGH OF SOUTHWARK

EFFRA ROAD SW.2

MILLSTREAM
ROAD SE1

BOURNE
STREET SW1
CITY OF WESTMINSTER

CANAL GROVE SE15
PRIVATE ROAD
LONDON BOROUGH OF SOUTHWARK

BROOK ROAD N.22

The Royal Borough of Kensington
and Chelsea
PONT
STREET, S.W.1

ELLFIELD AVENUE N.10.

London Borough of Hounslow
STAMFORD BROOK
AVENUE   W.6.

FLEET Rᴰ.

NECKINGER SE1
LONDON BOROUGH OF SOUTHWARK

WESTBOURNE
GROVE, W.11

Pudding Mill Lane

BRIXTON
WATER LANE SW2

CRANBOURNE ROAD N.10

London's Lost Rivers

rh
BOOKS

Published by Random House Books 2011

10 9 8 7

Copyright © Paul Talling 2011

Paul Talling has asserted his right under the Copyright, Designs and
Patents Act 1988 to be identified as the author of this work

First published in Great Britain in 2011 by
Random House Books
Random House, 20 Vauxhall Bridge Road,
London SW1V 2SA

www.rbooks.co.uk

Addresses for companies within the Random House Group Limited can be found at:
www.randomhouse.co.uk/offices.htm

The Random House Group Limited Reg. No. 954009

A CIP catalogue record for this book is available from the British Library

ISBN 9781847945976

Penguin Random House is committed to a sustainable future for our business, our readers and
our planet. This book is made from Forest Stewardship Council® certified paper.

Designed and typeset by Richard Marston
Maps by Darren Bennett
Photographs on p.16 and p.41 copyright © Thames Water
Printed and bound in China by C&C Offset Printing Co., Ltd.

River Effra

Contents

Introduction

In 2008, my five-year fascination with the run-down, closed-down buildings of London was rewarded with the publication of *Derelict London*, and I found with enormous relief that I was not alone. The book's website, derelictlondon.com, received so many hits and messages that I realised the capital's abandoned buildings were in good hands and began looking around for a new obsession. I found it in London's lost rivers.

Perhaps the most surprising thing about these hidden waterways is not that they have virtually disappeared into obscurity but that there are so many of them. I still find it mind-boggling to imagine that rivers ever poured down Farringdon Road and through the middle of Hackney, that vast merchant ships could sail straight across the Isle of Dogs and that Rotherhithe was once more water than land.

Writing this book has made me see London from an entirely new perspective: the city is riddled with watery relics and clues. Be it the name of an area (Muswell Hill), a tube station (Stamford Brook), a pub (The Falcon in Clapham), a road (Conduit Street) or just the sight of a mysterious lampless lamppost (see p.56), the evidence is abundant if you know what you're looking for. The 'lost' rivers, canals and docks of London really aren't as lost as they seem.

While many abandoned waterways gave way to railways and roads, the majority were culverted (covered over) and turned into sewers in an extraordinary feat of Victorian engineering. Londoners had been using the city's criss-crossing waterways for the disposal of human, household and industrial waste for centuries,

apid increase in population during the eighteenth and
th centuries had brought with it an increase in water-
ollution and disease. The final straw was the 'Great Stink'
he hot summer of 1858, when Parliament had to be relocated
e overwhelming smell of raw sewage.

Joseph Bazalgette
civil engineer charged with sorting out London's sewage
and saving everyone from the threat of cholera. His
achievement in designing a system of interconnecting
s to divert most of the city's waste away from the Thames.
these ageing sewers are now undergoing urgent
on, it is thanks to Bazalgette that London can cope with its
easing population, with only excess rainwater and what is
stically termed 'mixed water' trickling into the Thames after
ownpours.

I hope you enjoy walking the courses of these
ys, but I must advise great caution in exploring the Thames
e looking for the rivers' outfalls: ladders down to the Thames
ery precarious, while waste water and the incoming tide
heir own risks. Needless to say, it is extremely dangerous –
al – to enter any sort of sewer.

To send me any feedback or
ut about my occasional guided tours of London's lost rivers,
e on info@londonslostrivers.com.

River Westbourne

1
Major Rivers

River Westbourne

Best known nowadays for its grassy hills and desirable property, Hampstead was once most remarkable for its concentration of headwaters and rivulets, which splintered and converged in such a way as to supply central London with no fewer than three major rivers: the Westbourne, the Tyburn and the Fleet. In recent centuries, all three of these rivers have been covered over to make way for roads, railways and houses.

The Westbourne is the westernmost of Hampstead's rivers. Its upper stretch is made up of a number of minor streams that gradually join together, although the main source is generally agreed to have been a pond on Branch Hill. This pond was the subject of a handful of Constable paintings in around 1825, but the water gradually dried up towards the end of that century. On rainy days, tell-tale puddles in the sodden grass still reveal the river's source.

Marshy grassland is all that is left of Branch Hill Pond

1 Branch Hill
2 St Mary's Church
3 Site of Kilburn Wells
4 Regent's Canal
5 Bayard's Watering Place
6 Serpentine
7 Tyburn Brook
8 Chelsea Bridge
★ Station

Cricklewood

Willesden Green

WEST HAMPSTEAD

1 Hampstead

Swiss Cottage

Kilburn High Road

2

KILBURN

3

Kensal Rise

St John's Wood

Maida Vale

Ladbroke Grove

Bayswater

4

★ Paddington

5

7

HYDE PARK

KENSINGTON GARDENS

6

High Street Kensington

Knightsbridge

South Kensington

Victoria

Sloane Square

8

Pimlico

T h a m e s

Kilburn Abbey

In Kilburn, near the present-day junction of Kilburn Park Road and Shirland Road, the Hampstead streams converged with a smaller tributary from Queen's Park.

This section of the river was called the Kilburn or Kilbourne, from the Old English *cyne-burna* ('royal stream'), and it was a defining feature of the area. The small settlement that eventually grew into the Kilburn of today was established in the first half of the twelfth century, when a priory was built at the junction of the river and Watling Street, the ancient Roman road between Dover and Wales. The priory was supplied with fresh water and even fish by the Westbourne, and it became a popular stop-off for pilgrims on their way out of London. The location of the priory was more or less where Kilburn High Road station now stands and, although it was shut down during the Dissolution of the Monasteries under Henry VIII, its legacy is still visible in nearby road names: Kilburn Priory, Priory Road, Abbey Road and so on.

Springfield Lane and Springfield Walk, also near this junction, hint at another of the Westbourne's gifts to Kilburn: a spring whose waters were thought to have healing qualities. In 1714, the water from this spring was discovered to contain properties similar to Epsom salt, and for the next 100 years Kilburn was a fashionable spa. The former location of the Kilburn Wells is marked by a paving stone and a plaque at the corner of Kilburn High Road and Belsize Road, and the nearby Bell Tavern, demolished and rebuilt in 1863, plied a successful trade pumping spring water for 'the politest companies'.

A brass plaque in St Mary's Church, Priory Road

NEAR THIS PLACE LIE HUMAN REMAINS REMOVED FROM THE SITE OF KILBURN PRIORY 1857.

THIS FRAGMENT OF A BRASS SUPPOSED TO BE TO THE MEMORY OF A PRIORESS DATE ABOUT 1390 WAS FOUND 1877.

1770s KILBURN WELLS SPA

This paving stone is a reminder of Kilburn's fashionable past

Bayard's Watering Place and the Serpentine

The origin of the name Westbourne is unclear, and as late as 1824 the river was still marked on maps as Bayswater Brook. 'Bayswater' is thought to be a shortened version of 'Bayard's Watering Place', a name first recorded in 1380 in reference to a spot where bayards (horses) drank water from the river as it flowed under the Uxbridge Road (now Bayswater Road). The areas of West London that now have 'Westbourne' as part of their names – Westbourne Grove, for instance – are so named because they lay to the west of this bourne, or river.

Having crossed Bayswater Road at the end of Gloucester Terrace, the Westbourne's course leads straight into the Serpentine in Hyde Park. The river originally formed a series of natural ponds in the park, but they were linked up and dammed in 1730 on the orders of Queen Caroline, wife

Inside a Victorian sewer beneath Hyde Park

The parish boundary in the Italian Gardens

Beware Sloane Square and Knightsbridge

After leaving Hyde Park, the Westbourne followed a near-straight course all the way down to the Thames, forming a handy boundary between Westminster and Kensington and Chelsea. Street names provide some useful clues as to its former route – Bourne Street and Pont ('Bridge') Street in particular – but by far the most intriguing tales of the Westbourne come courtesy of its murky association with two of modern-day London's swankiest areas.

Knightsbridge – or Knight's Bridge – was a small settlement arranged around a stone crossing over the Westbourne. It lay outside the City of London and was quite the highwaymen's hot spot for centuries, as well as playing host to a military encounter between the citizens of London and Matilda, short-lived queen of England, in 1141. A little further downstream, the Westbourne passed beneath Blandel Bridge, later renamed Grosvenor Bridge but popularly

The Westbourne crosses Sloane Square station

The Westbourne's outfall near Chelsea Bridge

River Tyburn

As the lost rivers of London go, the Tyburn has left historians particularly perplexed. While its sources in North London and its uppermost stretches through Regent's Park have been relatively easy to chart, the Tyburn gets rather lost around Oxford Street and is almost anyone's guess as it passes through Westminster on its way to the Thames.

One source of the Tyburn was in the ancient manor of Belsize, a settlement that dates back to at least the early fourteenth century. The name Belsize comes from the French *bel assis*, 'beautifully situated'. The other source of the river is Shepherd's Well near the aptly named Spring Path in Hampstead, where a plaque remains to this day.

BOROUGH OF HAMPSTEAD

SPRING PATH N.W.3

FOR THE GOOD OF THE PUBLIC THIS FOUNTAIN IS ERECTED NEAR THE SITE OF AN ANCIENT CONDUIT KNOWN AS THE SHEPHERD'S WELL

HAMPSTEAD

BELSIZE PARK

Finchley Road ★

Swiss Cottage ★

Chalk Farm ★

St John's Wood ★

1 Regent's Canal

REGENT'S PARK

2 Regent's Park Boating Lake

Marylebone ★

Baker Street ★

Great Portland Street ★

3 Marylebone Lane

Marble Arch

Bond Street ★

4 Oxford Street

HYDE PARK

MAYFAIR

Piccadilly Circus ★

GREEN PARK

Hyde Park Corner ★

WESTMINSTER

5 Buckingham Palace

Victoria ★

7 Westminster Abbey

Pimlico ★

6 Vauxhall Bridge

Vauxhall ★

T h a m e s

1 Regent's Canal
2 Regent's Park Boating Lake
3 Marylebone Lane
4 Oxford Street
5 Buckingham Palace
6 Vauxhall Bridge
7 Westminster Abbey
★ Station

Regent's Park

The two sources of the Tyburn meet up just west of Primrose Hill and flow across Regent's Canal into Regent's Park. Unlike the Fleet, which was made to pass beneath the canal (see p.28), the Tyburn is carried across it by an aqueduct, the Charlbert Bridge. The only visible evidence of this footbridge's real purpose is an inspection hatch that pedestrians walk over on their way into the park.

Both the park and the canal were commissioned by the Prince Regent (later George IV), and both were developed in the early nineteenth century by architect John Nash. Most of Nash's grandiose plans never saw the light of day, but he did oversee the damming of the Tyburn into what is now the Regent's Park Boating Lake, which became one of the most popular features of the park when it was finally opened to the public in 1835. It achieved dubious notoriety in January 1867, however, when over 40 ice-skaters died after plunging through cracked ice. *The Times* solemnly reported the incident, casting polite aspersions on the park keepers, who, 'paying more regard to the necessities of the waterfowl than to the security of the skating public, broke the ice for some distance along the edges, thereby destroying the connexion of the central field with the shore'. A budding poet named Henry Disley was rather more sensationalist in his account: 'A poor faithful dog saw his master disappear,/ And never left the park since that evening,/ No food will he take, by the water stays near,/ For its master the poor dog is grieving.'

As a result of this disaster, the lake was drained and given a considerably shallower bed.

Charlbert Bridge carries the Tyburn over Regent's Canal

The Great Conduit and Mayfair

After leaving Regent's Park, the Tyburn meandered alongside what is now Marylebone Lane, which explains that road's curving line. There is some dispute as to whether the river followed Marylebone Lane all the way to Oxford Street – which was called Tyburn Road in the Middle Ages – with some eyewitness reports indicating the crossing was a little further west, where Oxford Street has a barely noticeable dip. The area around Marble Arch at the far end was once a village called Tyburn, home to the notorious Tyburn Tree gallows (see p.88).

Between 1236 and the Great Fire of London in 1666, a stone channel known as the Great Conduit carried Tyburn water from near Bond Street station – hence Brook Street and Conduit Street – all the way to Cheapside, which proved an effective way of maintaining water supply to the city's growing population. One theory for the increasingly mystifying route of the Tyburn below Oxford Street is that this conduit drew off so much water that the river's lower reaches were reduced to a near-insignificant trickle.

A plaque on Marylebone Lane recalls one of the Tyburn conduits

The Tyburn – complete with goldfish – in the basement of Grays Antiques

Grays Antiques on Davies Street staked a claim in London history in 1977 when, during renovation works, a stream supposedly identified as the Tyburn was discovered running through its basement. The river then continues along the eastern sides of Grosvenor Square and Berkeley Square, and exits Mayfair via Shepherd Market, where the infamous May Fair was held annually in the eighteenth century.

Palace through Victoria, and then followed King's Scholars Passage across the top of Vauxhall Bridge Road to Tachbrook Street, which took it down to the Thames. This stretch of the Tyburn formed part of the ancient boundary of Westminster Abbey's lands, and it is thought that the river's name means 'boundary stream'.

River Tyburn
Shepherds Well Belsize Park
Swiss Cottage Regents Park
Marylebone Oxford Street Mayfair
Green Park Westminster
Pimlico Tachbrook Street
River Thames

The Tyburn's outfall (far right) between Vauxhall Bridge and Battersea Power Station

A map of 1790 features the Tachbrook section as an open watercourse, although it was gradually downgraded to a sewer. In a dubious tribute to the King's Scholars of Westminster School, who are thought to have bathed and played in a nearby pond, this sewer is to this day called the King's Scholars' Pond Sewer.

The Tyburn's arched outfall into the Thames below Tyburn House can be seen to the west of Vauxhall Bridge. A plaque on the riverbank wall lists the river's route.

River Fleet

The River Fleet is the largest and best known of London's subterranean rivers, flowing from the heights of Hampstead Heath down to the Thames at Blackfriars Bridge. The Fleet was such a major river in Roman times that it culminated in a substantial estuary with a tide mill in it, and indeed the river takes its name from the Anglo-Saxon word *flēot*, meaning 'estuary'. The Fleet now exists as a large underground sewer.

The sources of the Fleet at Hampstead Heath are two springs separated by Parliament Hill, one on the western side near Hampstead and one on the eastern side in the grounds of Kenwood House. Some of the many rivulets that flow into the Fleet can still be seen on the Heath.

Hampstead Heath

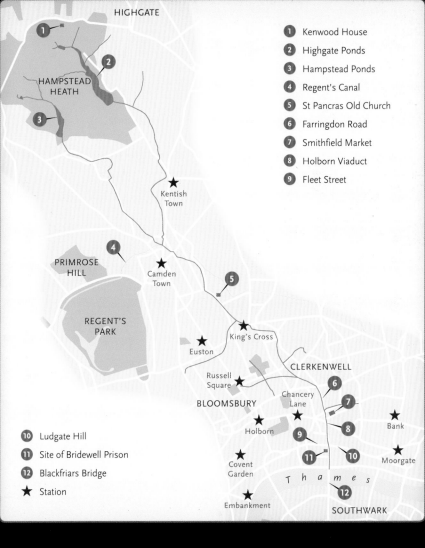

HIGHGATE

1 Kenwood House
2 Highgate Ponds
3 Hampstead Ponds
4 Regent's Canal
5 St Pancras Old Church
6 Farringdon Road
7 Smithfield Market
8 Holborn Viaduct
9 Fleet Street

HAMPSTEAD HEATH

Kentish Town

PRIMROSE HILL

Camden Town

REGENT'S PARK

Euston

King's Cross

CLERKENWELL

Russell Square

BLOOMSBURY

Chancery Lane

Holborn

Bank

10 Ludgate Hill
11 Site of Bridewell Prison
12 Blackfriars Bridge
★ Station

Covent Garden

Moorgate

Thames

Embankment

SOUTHWARK

down to Camden Town, where they meet up and continue
to King's Cross. This area was originally named Battle Bridge,
apparently in reference to an ancient river crossing where
Queen Boudicca fought in vain against the Romans in 60 AD.

Viaduct Pond, Hampstead

Men's Bathing Pond, Highgate

Legend has it that she is buried beneath platform 10 at King's Cross station. Battle Bridge became King's Cross in 1830, when an enormous statue of George IV was erected at the crossroads. This 'hideous monstrosity' (as historian Walter Thornbury described it in 1878) was removed after just 15 years.

The upper sections of the Fleet were covered when Hampstead was expanded in the 1870s, while the Camden and King's Cross sections were closed over during the development of the Regent's Canal from 1812 onwards. The Fleet now passes beneath the canal.

'River of Wells'

The most visible reminders of the Fleet's former prominence
in North London are the remains of wells that once dotted its
banks. These wells were reputed to have healing qualities
thanks to the water's high iron content, which explains why
they were so popular in the disease-ridden seventeenth
and eighteenth centuries, and why the Fleet earned the
nickname 'River of Wells'. The most elaborate example is the
Chalybeate Well at Well Walk, Hampstead, which was opened
for the benefit of the poor in 1698.

The Bagnigge Wells near King's Cross
were opened as a spa in 1760 and soon achieved notoriety
as a location for illicit rendezvous. Appropriately enough,
Charles II's mistress Nell Gwynne lived at the nearby Bagnigge

LEFT Chalybeate Well; RIGHT Bagnigge House

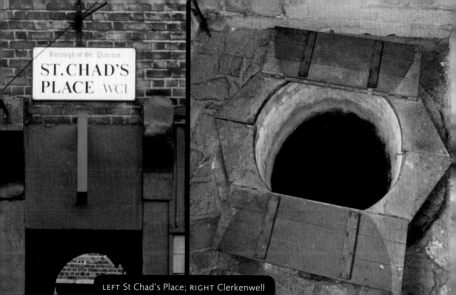

LEFT St Chad's Place; RIGHT Clerkenwell

The Fleet and the Railway

The Fleet ran beside the church of St Pancras, which stands
near the present-day international railway station of the same
name. St Pancras Old Church is one of Europe's most ancient
sites of Christian worship and may even date back as far as the
early fourth century, although the building itself has undergone
substantial renovation since then. A board on the church's
railings depicts bathers in the Fleet in 1827.

In 1865, a trainee
architect named Thomas Hardy, who later became rather
more famous as a novelist and poet, was overseeing the careful
removal of bodies and tombs from the part of the churchyard
on which the Midland Railway line was being built. Instead of
destroying the headstones, Hardy positioned them around an
ash tree – now known as the Hardy Tree – where they can still
be seen today.

After King's Cross and St Pancras, the Fleet's route is
easy to chart on any London A-Z. It follows the line of King's
Cross Road and then flows all the way down Farringdon Road
to the Thames. The construction of the Metropolitan Line in
1862 buried the Fleet under Farringdon Road (you can still hear
the water through a grating in front of the Coach and Horses
pub on Ray Street), although the river created problems later
that year when the sewer burst, causing a number of the arches
lining the tunnel to collapse.

Over 100 years later, the Fleet almost
gave its name to another London Underground line, but since
its opening coincided with Queen Elizabeth II's silver jubilee in
1977, the proposed Fleet Line was renamed the Jubilee Line.

The Hardy Tree

Holborn and Smithfield

The stretch of the Fleet that flowed down the valley of what is now Farringdon Road was called the Holbourne (or Oldbourne), from the word *holburna* ('hollow stream'), a reference to the river's deep valley – in some places it is almost 25 feet below street level. The Holborn Viaduct, a large iron bridge opened by Queen Victoria in 1869, spans what is known as the Fleet Valley.

The construction of Farringdon Road in the 1840s and 1850s – an impressive feat of engineering in its day – had not only buried the Fleet but had also destroyed Fleet Market, a collection of shops and stalls erected in the 1730s on a covered section of the river, next to where Smithfield Market stands today. It also carved through the riverside quarter of Hockley-in-the-Hole (Ray Street), which was notorious for bear-baiting and dog-fighting.

The Fleet Valley as seen from Holborn Viaduct, overlooking Smithfield Market

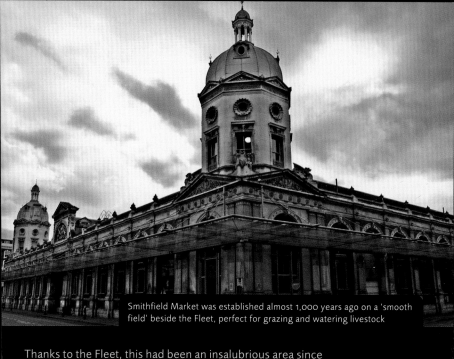
Smithfield Market was established almost 1,000 years ago on a 'smooth field' beside the Fleet, perfect for grazing and watering livestock

Thanks to the Fleet, this had been an insalubrious area since at least the thirteenth century. As London grew in population, the human and industrial waste pouring into the water – not least the remains of carcasses from Smithfield Market – began to clog the once-great river. The lower section of the Fleet became known as the Fleet Ditch, and for centuries it was little more than a vast open sewer. In 1710, Jonathan Swift (author of *Gulliver's Travels*) depicted the filthy Fleet in his poem 'A Description of a City Shower': 'Sweepings from Butchers Stalls, Dung, Guts and Blood, / Drown'd Puppies, stinking Sprats, all drench'd in Mud, / Dead Cats and Turnip-Tops come tumbling down the Flood.'

Prisons and Plague

The Fleet Valley was still notorious for its slum dwellings, bad characters and diseases in 1838, when Charles Dickens used it for the location of Fagin's den in his novel *Oliver Twist*. Indeed, between the twelfth and nineteenth centuries, the increasingly undesirable and therefore cheap land of the Fleet Valley was a popular spot for building prisons, including Newgate, Clerkenwell, Ludgate, Fleet and Bridewell. The introduction of cholera into Clerkenwell Prison in 1832 was attributed to the effluvia of the river.

Fleet Prison was situated north of Ludgate Hill, behind what is now City Thameslink station. Until its closure in 1844, the prison mainly housed debtors and bankrupts. Just below it, the Fleet flowed between the end of Fleet Street and

Ludgate Circus

FLEET
STREET EC4

POETIC
PW3

The End of the Fleet

As well as wiping out the Plague of 1665, the Great Fire in 1666 served as a catalyst for much-needed architectural and sanitation improvements all over London. Extensive building work took place across the city, including the construction of the present St Paul's Cathedral by Christopher Wren.

Wren was also charged with converting the lower reaches of the Fleet – which had already been cleaned twice that century due to obstruction by offal and sewage – into a usable asset. The New Canal, as it became known, was based on the elaborate Grand Canal in Venice. The mouth of the Fleet was broadened to a width of forty feet and flanked with great wharves for unloading coal from the north of England (hence Newcastle Close off Farringdon Street), but the torrent of pollution from upriver, which was still basically an open sewer, caused the canal to be a failure. The apparent impossibility of convincing Londoners to dispose of their detritus elsewhere was compounded by the unmanageable rate at which the river was silting up.

The Fleet became so choked that it was no longer navigable, and after several people had fallen to their deaths in the mire, the canal – the last open stretch of the Fleet – was gradually covered over. The section above Ludgate Circus was closed in 1737 and the lower section in 1769.

The Fleet now spills into the Thames from an unassuming arch hidden beneath Blackfriars Bridge.

Inside the Fleet sewer

River Walbrook

Despite being one of the shortest rivers ever to flow through London, the Walbrook tends to be counted as a major waterway since it carved a route right through the historic City of London.

Surprisingly for such a short river, the exact location of the Walbrook's source has been a matter of great dispute, although evidence suggests that it was somewhere near the junction of Curtain Road and Holywell Lane in Shoreditch. The latter was the site of a holy well that is thought to indicate the source of the Walbrook or the Black Ditch (see p.128) or both. 'Shoreditch' itself may be derived from 'sewer ditch' in reference to the Walbrook, a plausible theory considering the widespread pollution of London's historic waterways.

Holywell Lane, Shoreditch

1 Holywell Lane
2 Barbican
3 London Wall
4 All Hallows-on-the-Wall
5 Houndsditch
6 Bank of England
7 Walbrook
8 Temple of Mithras
9 London Stone
10 Southwark Bridge
11 London Bridge
★ Station

Hoxton

SHOREDITCH

Old Street

Shoreditch High Street

Moorgate

Liverpool Street

St Paul's

Bank

CITY OF LONDON

Cannon Street

Monument

Aldgate

Fenchurch Street

Tower Hill

SOUTH BANK

Thames

London Bridge

Southwark

SOUTHWARK

London Wall

When the Romans established the settlement of Londinium –
today's City of London, more or less – nearly 2,000 years ago, the
Walbrook most likely influenced their choice of location. It wasn't
until the late second century that they marked the boundaries
of Londinium with an enormous wall, but what became clear
once they did was that the epicentre of their city was where the
Walbrook formed a slightly wonky right-angle with the Thames
near Cannon Street station. It was an ideal transport link, water
supply and sewer running straight through the middle of the
settlement. The wall itself remained largely intact into the
eighteenth century.

The name Walbrook is probably derived from the river's
route through London Wall, chunks of which can still be seen
outside the Museum of London, at the Barbican Estate and in

A section of London Wall at the Museum of London

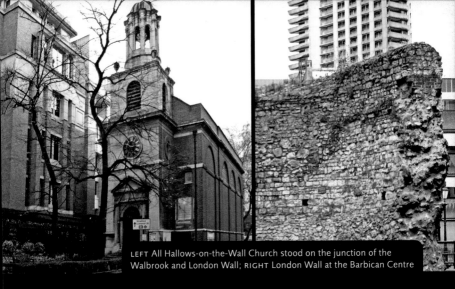

LEFT All Hallows-on-the-Wall Church stood on the junction of the Walbrook and London Wall; RIGHT London Wall at the Barbican Centre

front of Tower Hill station. The river approached the wall via what is now the Broadgate development west of Liverpool Street station – where it met a tributary from the Aldgate end of Middlesex Street – and Blomfield Street, where it received a second tributary that originated under the Barbican Centre and ran across the top of Finsbury Circus.

The Walbrook crossed the wall just west of the church of All Hallows-on-the-Wall, which was constructed in the 1760s on the site of a twelfth-century original and now stands on the road confusingly named London Wall. A swampy defensive ditch that surrounded the city in the Middle Ages was fed by water from the Walbrook, with the eastern stretch of it supposedly nicknamed Houndsditch as early as the 1400s because it was a suitably murky place for dumping dead dogs and other waste.

Walbrook and the Temple of Mithras

Once inside the walled city, the Walbrook flowed in a south-westerly direction towards what is now the Bank of England, under which it passes. The church of St Margaret Lothbury, just behind the Bank, was built on a culverted section of the river in the thirteenth century and rebuilt by Christopher Wren after the Great Fire of 1666. Below Cheapside, the Walbrook ran alongside the road now named Walbrook, and it is here that some of the river's most fascinating history is to be found.

Construction work beneath Walbrook in 1954 revealed the ruins of a Roman temple thought to have been built beside the river in the third century. The haul of artefacts recovered from the site led archaeologists to conclude that the temple had been dedicated to Mithras, a Persian god associated with a cultish religious fellowship that was popular among the Roman military in the days before Christianity became prominent. Worshippers of Mithras frequently depicted him slaughtering a sacred bull in a cave – their underground temples symbolised this cave – and they may even have bathed in bulls' blood in the hope of enhancing their own virility.

The remains of the Temple of Mithras were relocated to Temple Court on Queen Victoria Street, while the various artefacts are on display at the Museum of London. There are plans afoot, led by British architect Norman Foster, to restore the temple to its original location.

The remains of the Temple of Mithras on Queen Victoria Street

Also on Walbrook is the church of St Stephen Walbrook, which originally stood on the western bank of the river but was reconstructed on the eastern side in the 1430s, and then rebuilt again after the Great Fire by Christopher Wren.

London Stone and Cannon Street

The Walbrook, like most of London's lost rivers, now exists as a sewer, but some lower sections remained open into the late eighteenth century. The historian John Stow in his *Survey of London* (1598) recalled that the water had 'such a swift course that in the year 1574 a lad of eighteen years minding to have leapt over the channel, was borne down that narrow stream towards the Thames with such violent swiftness as no man could rescue or stay him'.

The river crosses Cannon Street just west of Walbrook, passing very close to another of the city's little-known Roman artefacts, the London Stone. This stone – supposedly the original – was used by the Romans to indicate the epicentre of Roman Britain, and to measure any distances into and out of London. (Charing Cross is used for the same purpose these days.) It gradually became the traditional place for swearing

The London Stone sits in this neglected nook on Cannon Street

Walbrook Wharf with Southwark Bridge in the background

oaths and making proclamations, and the rebel Jack Cade is said to have struck his sword on it and declared himself mayor during his rebellion against Henry VI in 1450, a scene immortalised by Shakespeare in his play *Henry VI, Part 2* (1591). The stone can be found set into the wall of 111 Cannon Street, behind a grille and a grubby sheet of glass.

Below Cannon Street, the Walbrook made a sharp turn to the east at a road formerly (and aptly) known as Elbow Lane, which is now College Street, before meeting the Thames at an inlet named Dowgate. The outfall of the Walbrook sewer is now called Walbrook Wharf, a small dock beside Cannon Street station from which container vessels carry the city's waste up the Thames to Essex.

River Effra

Despite being among the largest of London's lost rivers, twelve feet wide in parts and with a name thought to be derived from the Celtic word for 'torrent', the Effra starts out in so un-torrentlike a fashion that its sources and early route are unclear.

The principal source of the Effra is somewhere in the hills of Norwood – which is named after the Great North Wood that covered most of the area with dense oak trees – most likely at the Harold Road end of what is now Upper Norwood Recreation Ground. Like many of London's lost rivers, the Effra was turned into an underground sewer in the mid-nineteenth century, and gurgling noises emanating from a drain cover off Hancock Road are thought to be the sounds of it flowing underneath.

The source of the Effra?

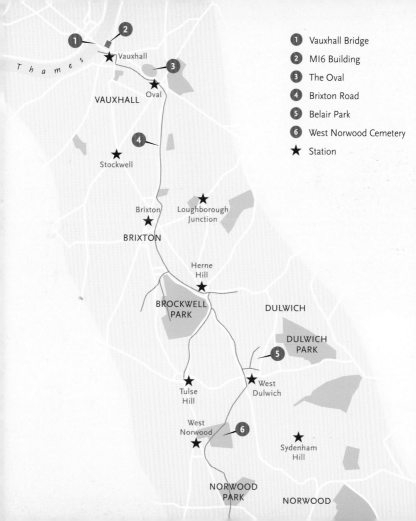

1 Vauxhall Bridge
2 MI6 Building
3 The Oval
4 Brixton Road
5 Belair Park
6 West Norwood Cemetery
★ Station

Thames

Vauxhall

VAUXHALL

Oval

Stockwell

Brixton

Loughborough
Junction

BRIXTON

Herne
Hill

BROCKWELL
PARK

DULWICH

DULWICH
PARK

Tulse
Hill

West
Dulwich

West
Norwood

Sydenham
Hill

NORWOOD
PARK

NORWOOD

Great Floods

The Effra's elusive nature has not prevented it from making frequent appearances in the form of flash floods. In 1890, much of the brick wall surrounding the House of the Congregation of Our Lady of Fidelity – now Virgo Fidelis Convent School – was swept away when the Effra flooded West Norwood, and the repairs are still visible in the existing wall.

As indicated on a map of 1800 and supported by the natural contours of the area, the Effra seems originally to have flowed through West Norwood Cemetery (one of London's 'Magnificent Seven' cemeteries – see also p.72 and p.160) on its way towards the valley of Croxted Road. An unconfirmed urban myth holds that a coffin from an undisturbed grave in West Norwood Cemetery was found floating in the Thames in the nineteenth century, having subsided into the underground Effra and sailed downstream to Vauxhall.

In 1914, a three-hour storm caused the sewer to overflow once more, flooding houses and forcing residents to evacuate the area. In keeping with the religious theme of Norwood's floods, the story goes that a number of families' Sunday roasts were washed into the streets around Pilgrim Hill. The aptly named Boat House, which would never actually have stood at the river's edge, was one of the buildings worst affected by the flood.

Further floods in the 1920s and 1930s forced the authorities to widen the sewer, but this proved only a temporary fix as the area was flooded once again during a torrential downpour in 2007.

LOVING
...WH...
...G...N
WHO PA...
NOVEMBE...
...D 8...Y...
...NT...TH...TY...
...ARLES EDWAR...
HEATH
...ED FEB 4TH 195...
...GED 84 YEARS

West Norwood Cemetery

Belair and Brockwell

One of the last visible vestiges of the Effra is at Belair Park, on the Dulwich side of Croxted Road. Built in 1785, the original mansion was set in grounds that featured an ornamental pond created by damming a section of the river. The pond has become a haven for wildlife and has the dubious honour of being one of the only places in London where rats will go about their daily business unperturbed by human observers.

As the Effra reached what is now Herne Hill station, it was joined by two further tributaries, one that ran down Norwood Road and a shorter one from Half Moon Lane. This latter tributary was described by John Ruskin, the nineteenth-century painter and critic who hailed from Herne Hill, as a 'tadpole-haunted ditch'. He once claimed that his first sketch of any merit was of a bridge over the Effra at the foot of Herne Hill, more or less where the Half Moon Tavern stands today. The river was covered over in this area in the 1820s.

The river flowed along what is now Dulwich Road (formerly Water Lane) and was joined at the crossroads of Effra Parade and Brixton Water Lane by another tributary from Brockwell Park. Nearby Effra Road indicates the boundary of an agricultural area known as Effra Farm rather than the course of the river, which ran through the farmland.

The source of one tributary of the Effra can be seen in Brockwell Park

Brixton and Vauxhall

Effra Farm was the name used around the turn of the nineteenth
century for the chunk of land to the east of Effra Road and below
Coldharbour Lane. The farmland was situated in the Manor
of Heathrow – nothing to do with the airport – and there has
even been tenuous speculation that the name Effra is a London
bastardisation of 'Heathrow'.

The course of the river is relatively easy to follow
from Brixton to the Thames at Vauxhall. A number of Victorian
'stink pipes', lamppost-like tubes that lifted the overpoweringly
awful sewer stench away from street level, line the route, including
ones along Dulwich Road at the junctions with Chaucer Road
and Brixton Water Lane. Then, from Brixton station onwards,

The outfall of the Effra beneath MI6 in Vauxhall

the Effra ran along the eastern side of Brixton Road all the way to the Oval, which partly owes its distinctive shape to the curve of the river just south of it, and which suffered minor flood damage in 1950. London lore would have it that Cnut the Great sailed up the Effra to Brixton during the 1016 Danish conquest of England, a slightly more plausible story than that of Elizabeth I travelling all the way to Dulwich by barge.

The final stretch of the Effra was once known as Vauxhall Creek, a sizeable channel that ran through what are now the St George Wharf development and Vauxhall station. There is still a small outfall beneath St George Wharf, although a larger outlet – a more recent storm relief channel, complete with a sign saying 'Effra' – can be seen on the other side of the bridge, directly beneath the MI6 building.

Bollo Brook

2

Waterways of West London

BOLLO BROOK

STAMFORD BROOK

PARR'S DITCH

COUNTER'S CREEK

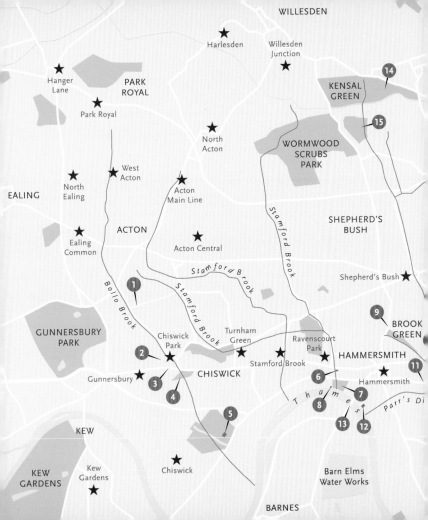

tting Hill Gate

KENSINGTON

High Street Kensington

16

17 EARL'S COURT

South Kensington

CHELSEA

Counter's Creek

Fulham Broadway

18

19

Thames

West Brompton

Bollo Brook

Bollo Brook rises near Ealing Common station and more or less follows the Ealing branch of the District Line – or indeed vice versa – before splitting away in Chiswick and flowing down to the Thames opposite Barnes.

The brook's unusual name is thought to come from a bridge originally named Bull Hollow Bridge but which over time became known as Bollo Bridge, a structure that no longer exists except in the name of Bollo Bridge Road in Acton. Between Acton Town and Chiswick Park stations, the Bollo followed the meandering line of Bollo Lane – to this day the boundary between the boroughs of Hounslow and Ealing – passing the site of a modern-day gastropub also called the Bollo. Some of the natural water features of nearby Gunnersbury Triangle Nature Reserve are thought to be fed by the brook.

Bollo Brook approached Chiswick High Road along the lower section of Acton Lane, and then carved a line straight across Turnham Green, towards what is now Chiswick Town Hall. This small public park was the site of a major stand-off between the Cavaliers (Royalists) and Roundheads (Parliamentarians) in November 1642, during the First Civil War. Charles I's army failed miserably in its aim of winning local support and storming triumphantly into London, and instead had to beat a retreat back to Oxford.

Below Turnham Green, Bollo Brook flowed southwards for a short while before curving round into the grounds of Chiswick House. The name may well be coincidental, but Eastbourne Road just below the A4 suggests that the brook passed alongside some or all of that street.

Gunnersbury Triangle Nature Reserve

Chiswick Park station stands on a hill that once overlooked Bollo Brook

Chiswick House

After a relatively uninspiring route under Acton and residential Chiswick, Bollo Brook suddenly emerges as part of England's first landscaped garden, in the grounds of Chiswick House. This neo-Palladian mansion lies just west of the Hogarth Roundabout, so named in dubious tribute to William Hogarth, the eighteenth-century satirist and painter of *A Rake's Progress*, whose nearby peaceful country retreat is now hemmed in by the A4.

What is now known as Chiswick House was designed and built as a country villa by the third Earl of Burlington in the 1720s. Bollo Brook originally formed the boundary of the Burlington estate, but when the property was extended by the purchase of land on the other side of the stream, Bollo Brook was widened and canalised and the adjacent grounds

This elegant stone bridge was built in 1774 to replace the original wooden bridge over Bollo Brook

Bollo Brook is now carried in a pipe beneath the Chiswick House lake

extensively landscaped. Although the villa was originally
intended as a storage depot for the earl's art and furniture, it
ultimately took on the name Chiswick House when the original
mansion was demolished in the late eighteenth century.

The Bollo once
fed the lakes and fountains at Chiswick House, but its use as a
waste-water sewer by laundries upstream meant that it had to be
culverted into a pipe beneath the lake in order to preserve the
gardens' beauty. A tithe map of 1846 indicates that the Chiswick
Park lake continued all the way to the Thames, but since 1936 any
surplus water from the gardens has joined Bollo Brook beneath
the A316, continuing underground to the distinctly unattractive
concrete outfall at the end of Promenade Approach Road.

Stamford Brook

Stamford Brook is made up of three converging streams that only flow as a unified waterway on their final approach to the Thames at Hammersmith.

The westernmost stream, which originates in the vicinity of Mill Hill Road in Acton, has been the subject of some debate as there is evidence to suggest it might have been a tributary of Bollo Brook. The high turnover of both natural and artificial waterways in this part of West London has made exact routes difficult to pin down, but local archives state that the disputed stream is in fact Mill Hill Brook, and as such a tributary of Stamford Brook. It flows towards Chiswick Common and then runs north of the District Line track, past Stamford Brook station, to Ravenscourt Park.

One branch of Stamford Brook helped plot a route for the District Line between Chiswick and Hammersmith

A fashionable spa was once situated by Wormwood Scrubs

The middle stream, the Warple, rises near Acton Main Line station and flows down Horn Lane before veering off to the east and picking a route through residential streets towards the junction of Goldhawk Road and Stamford Brook Road, where it meets Mill Hill Brook. The eastern stream comes from just north of Wormwood Scrubs, where the fashionable Old Acton Wells once stood. Their mildly laxative waters were sold in casks in central London during the eighteenth century, but the only remaining trace is the name of Wells House Road. This stream follows Common Lane, Old Oak Road and Askew Road down to Ravenscourt Park, a stretch of water that became so polluted by the late nineteenth century as to be described as a 'disgraceful open sewer'.

Appropriately enough, much of Stamford Brook had been converted into sewage pipes by the turn of the twentieth century.

Little Wapping, Hammersmith

At the confluence of Mill Hill Brook and the Warple, a secondary ditch carried some of Stamford Brook's water south along British Grove to the Thames, while the main stream cut through the middle of Ravenscourt Park. Here it formed a moat around the manor house of Palingswick or Paddenswick, which was destroyed by bombing during the Second World War. Just beyond the park, this stream converged with the one from Wormwood Scrubs and the unified Stamford Brook flowed down Paddenswick Road and Dalling Road to King Street.

The name Stamford comes from an ancient 'stony ford' that crossed the brook at King Street in front of today's town hall, which stands on the site of the former Cromwell Brewery. The Stamford Brook area was known for its laundries and breweries, which made good use of the streams in times past.

Stamford Brook became wider between King Street and the Thames, and was until the early nineteenth century a navigable watercourse known as Hammersmith Creek. The surrounding industrial area featured lead mills, malt houses and ship builders, as a result of which it was nicknamed Little Wapping. Barges used to sail up to the brewery past the seventeenth-century Dove pub, which still stands. The pub's other claims to fame are the shortest bar in the country and the upstairs room in which 'Rule, Britannia!' was supposedly written.

The Creek was filled in in 1936, and a flying bomb devastated the last remnants of Little Wapping in 1944. Furnival Gardens was created on the site in time for the 1951 Festival of Britain, and Stamford Brook's outfall into the Thames can be seen below it.

A stink pipe on the corner of Chiswick Mall and Eyot Gardens, along the course of Stamford Brook's secondary ditch

Parr's Ditch

The U-shaped Parr's Ditch, which is thought to have been a man-made boundary between Hammersmith and Fulham, originates at the western end of Brook Green. It most likely served a second purpose as an open sewer, contradicting the theory that salmon ('parrs') might ever have been found swimming in it, and it was converted into an official sewer in 1876. A stink pipe (see p.56) on the patch of grass by the Brook Green Hotel indicates where the ditch may have started.

Although there is evidence of human occupation in Hammersmith going back to Roman times and possibly beyond, Brook Green itself was established as a small hamlet in around the seventeenth century. By the eighteenth century, it was known for two

This pathway lined by ancient trees marks the course of Parr's Ditch along Brook Green

A boundary marker between the historic parishes of Hammersmith and Fulham at the Parr's Ditch outfall by the Riverside Studios

things: a predominance of Catholic institutions, thanks to which it was nicknamed Pope's Corner, and a series of market gardens on either side of the 'brook'.

Parr's Ditch ran the length of Brook Green before curving down across Hammersmith Road and Talgarth Road – where there is another stink pipe – and looping back around Hammersmith Cemetery towards the Thames at Riverside Studios. On its way out of Brook Green it would have passed a pub called the Black Bull Inn – where a terracotta former school building now stands at 153 Hammersmith Road – which accounts for the ditch's alternative name, Black Bull Ditch.

It seems probable that a later ditch was dug between the Brook Green Hotel end of Parr's Ditch and the confluence of Stamford Brook's various streams at Ravenscourt Park. This theory is backed up by the remarkable number of stink pipes along the route this second ditch would have taken, particularly along Aldensley Road.

Counter's Creek

Counter's Creek rises on the northern side of Kensal Green Cemetery and flows in a largely straight line down to the Thames at Chelsea Creek. As a natural watercourse that has been canalised, sewerised, converted into train tracks *and* used as a borough boundary, Counter's Creek is the archetypal lost river.

After passing through Kensal Green Cemetery – one of London's 'Magnificent Seven' alongside West Norwood (see p.52) and Abney Park (see p.160) – Counter's Creek ran down the eastern side of Little Wormwood Scrubs, where it was later converted into ornamental ponds before being filled in. Today's concrete footpath is a replacement for the gravel walkway that originally bordered the water.

Counter's Creek historically acted as a boundary between settlements, a service it still provides to this day between the boroughs of Kensington and Chelsea and Hammersmith and Fulham. An attempt to canalise the lower section in the late 1820s (see overleaf) was unsuccessful, as was a railway line constructed over the creek's course between Willesden and the canal basin in Kensington. In 1863, however, the railway was extended over almost the entire length of Counter's Creek, and this route still exists as part of the London Overground network between Shepherd's Bush and Imperial Wharf stations.

Kensal Green Cemetery

The creek's name is thought to derive from Counter's Bridge, which traversed the water near Olympia. This bridge has been traced back as far as the fourteenth century, when it was referred to as Contessesbregge, possibly in reference to the Countess of Oxford, who owned the land at the time. Stamford Bridge (no connection to Stamford Brook) was formerly known as Sandford Bridge, 'the bridge at the sandy ford'.

Kensington Canal

The main use of Counter's Creek in the days before industrialisation was as a backwater for carrying sewage down to the Thames. There are no records suggesting that it was navigable until the lower two miles of it were turned into the 100-foot-wide Kensington Canal in 1828. The canal basin was just below Counter's Bridge, where the junction of Kensington High Street and Warwick Road is today.

Kensington Canal proved to be a financial failure and it was bought up in the middle of the nineteenth century by the West London Railway Company, which converted the water into a sewer and extended the existing railway track from Willesden down to the King's Road. This railway was connected with Clapham Junction station on the other side of the Thames in 1863.

The last stretch of Kensington Canal, Chelsea Creek, branches away from the railway line and remained a working canal – albeit a very short one – for barges delivering coal to the power station on Lots Road, which provided electricity for the London Underground system. This commercial use of the canal came to an end in the 1960s, when Lots Road Power Station converted from coal to oil. A neglected section of the canal can still be seen from Lots Road Bridge.

There is further evidence of the former canal beside the westbound platform of West Brompton station, where a rubbish-strewn ditch floods during heavy rain. The remains of a bridge over the canal can also be spotted adjacent to this station, in the shadow of Earl's Court Exhibition Centre.

Counter's Creek running alongside Lots Road Power Station

Grosvenor Canal

3
Waterways of Central London

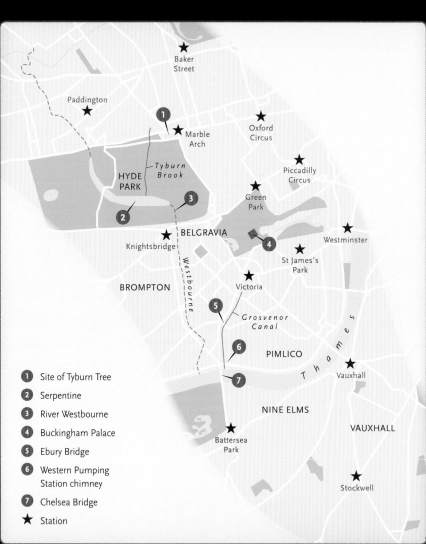

1 Site of Tyburn Tree
2 Serpentine
3 River Westbourne
4 Buckingham Palace
5 Ebury Bridge
6 Western Pumping Station chimney
7 Chelsea Bridge
★ Station

PRIMROSE HILL

★ Camden Town

Cumberland Arm

REGENT'S PARK

★ King's Cross

★ Euston

BLOOMSBURY

★ Baker Street

★ Regent's Park

★ Warren Street

Tottenham Court Road

★ Holborn

★ Bond Street

Oxford Circus

SOHO

★ Covent Garden

Cock and Pye Ditch

Embankment

★ Green Park

★ Waterloo

★ Westminster

St James's Park

Thames

LAMBETH

1 Regent's Canal

2 London Zoo

3 Gloucester Gate Bridge

4 Regent's Park Barracks

5 Site of Cumberland Market

6 Seven Dials

7 St Martin's Lane

8 St Martin-in-the-Fields

9 Victoria Embankment

10 Hungerford Bridge

★ Station

Grosvenor Canal

The shortest canal in London started at the Thames in Pimlico,
just east of Chelsea Bridge, and originally went as far as its
basin, half a mile north on the site of Victoria station. It was
begun in 1725 as a navigable tidal channel providing access
to the nearby Chelsea Waterworks Company, which supplied
Thames and Westbourne water to much of central London in
the eighteenth and nineteenth centuries.

The Chelsea Waterworks
Company was quite the industrial marvel during its 100 years
of prominence. By the mid-eighteenth century, it was using
an improved version of the Newcomen steam engine to pump
water from the river, apparently to the great displeasure of
the Buckinghams of Buckingham Palace when the wind was
blowing in their direction. Around this time, the same company
became the first to use iron pipes rather than leaky wooden
ones, and it also introduced a novel system of slow sand
filtration in the 1820s, after being censured for the impurity
of its water.

In 1856, however, four years after an Act of Parliament
banned the use of metropolitan Thames water for household
use, the Chelsea Waterworks Company was forced to relocate
to Surbiton. A remaining waterworks building, the Western
Pumping Station, still stands beside the site of the canal and
the boiler chimney is something of a local landmark, despite
the fact that it nowadays acts as a sewage ventilation shaft.

The entrance to Grosvenor Canal from the Thames beneath Grosvenor Road, with the Western Pumping Station chimney in the background

The Shrinking Canal

The Chelsea Waterworks Company inlet was extended into a canal and opened to traffic in the 1820s, taking its name from the Grosvenor family, historic Dukes of Westminster. This canal extension was designed for carrying goods and building materials to and from Belgravia, which was being developed at the time. Wharves receiving timber, coal and stone were situated around the canal basin.

When the waterworks moved out of the area in the mid-nineteenth century, the vacated land was used by various railway companies building lines into West London. The Grosvenor Canal basin was filled in to make way for Victoria station, which opened in 1860, and the canal was shortened so as to terminate at Belgrave Road. It was further shortened to end at Ebury Bridge when the station was enlarged in 1899.

A plaque commemorating one of Grosvenor Canal's many transformations

The bricked-up arch of the canal under Ebury Bridge

Westminster City Council bought the truncated Grosvenor Canal in 1905 and in the late 1920s shortened it yet again in order to accommodate the Ebury Bridge Estate. The bricked-up arch beneath Ebury Bridge is the last remaining indication that the canal ever extended further north.

For the remainder of the twentieth century, Grosvenor Canal was used for refuse disposal, with bargefuls of rubbish from Westminster and beyond being transported along the Thames out of London. In 2000, permission was granted for the short stretch of canal to be incorporated into the new Grosvenor Waterside residential development, as an ornamental water feature running along the canal's original route.

The Cumberland Arm

The Cumberland Arm, also known as the Cumberland Market Branch, was a short stretch of the Regent's Canal that branched off at the eastern end of Prince Albert Road and ran parallel to Albany Street until it reached its basin at Cumberland Market, near Euston station.

The Cumberland Arm was developed in the 1810s at the same time as the main Regent's Canal (see p.22), and remained in active service until the Second World War, after which it was filled in and the land used for housing and roads. The fork where the Cumberland Arm branched off from the Camden Lock route of the canal can still be seen today, although it is now a dead end, home to a small number of moored barges and, perhaps unexpectedly, an enormous floating Chinese restaurant.

The route to the left takes barges up to Camden Lock, while the dead-end on the right is the only remnant of the Cumberland Arm

These crumbling faces once watched over traffic on the Cumberland Arm

The Cumberland Arm's route took it straight through what is now the car park for London Zoo, after which it flowed beneath an ornamental bridge on its way southwards. Gloucester Gate Bridge, as it continues to be known, remains intact and anyone driving along Prince Albert Road could be forgiven for assuming there is still water running beneath it. In reality, the view from one side is of the aforementioned car park, and from the other a series of gardens belonging to the exclusive houses of Park Village East, which were designed as canalside properties by John Nash, the architect of Regent's Park. The bridge itself, though rather falling into disrepair, still shows vestiges of its former glory: it features an ornate lamp at each corner and, above the infilled arches on the 'water' side, a series of crumbling motifs.

Albany Street and Cumberland Market

Albany Street runs southwards along the eastern side of
Regent's Park and was named after the Duke of Albany, younger
brother of the Prince Regent (later George IV). The Cumberland
Arm followed it all the way down to its basin, and the road's
two most prominent features – Regent's Park Barracks and
Cumberland Market – were the major causes of traffic along this
short stretch of water.

The barracks were constructed during the rapid
development of the Regent's Park area and later housed the
Royal Horse Guards between the 1890s and the 1960s. Barges
brought supplies for both the military and their horses, including
ammunition from Woolwich Arsenal (see p.136).

Cumberland Market,
a little further south, was built in the late 1820s as a new home
for the hay market – literally a market selling hay and straw for
horses – which gave its name to the road off Piccadilly. The new
market also sold agricultural produce and traded in livestock, all
of which were transported into central London by barge. The
market continued in an ever-decreasing capacity for 100 years,
although the area began its steady decline after a devastating
outbreak of cholera during the 1850s.

After the Second World War,
the largely disused canal basin was filled in with rubble from
destroyed buildings, and its wharves and warehouses were
replaced with council housing. The area that once bustled with
market activity is now a series of allotments, whose owners –
urban legend would have it – occasionally uncover fragments
of old barges beneath their plots.

The Cumberland Arm ran along the back of Regent's Park Barracks

The basin at Cumberland Market was filled in with wartime rubble and is now used as allotments

Tyburn Brook

Although it shares its names with one of London's major lost rivers (see p.20), Tyburn Brook is in fact a tributary of the Westbourne (see p.12). The very short brook ran from the vicinity of Marble Arch to the Serpentine in Hyde Park.

Although very little is known about Tyburn Brook and there is no real evidence of its exact route, the origin of its name is much documented. During the Middle Ages, the crossroads on which Marble Arch now stands was the junction of Tyburn Road and Tyburn Lane (Oxford Street and Park Lane), and there was a village here called Tyburn. Confusingly, the names of this settlement and its associated roads most likely come from the nearby River Tyburn, whose name means 'boundary stream'.

Between the twelfth and eighteenth centuries, the village of Tyburn was most famous (or infamous) for its much-used gallows, which was upgraded to a newfangled multi-execution contraption in the sixteenth century and thereafter nicknamed the Tyburn Tree. People in their tens of thousands would crowd around the Tyburn Tree to watch criminals and heretics be put to death. Notable executions on this spot include those of Roger Mortimer (1330), who deposed Edward II with help from the queen; Elizabeth Barton (1534), a Catholic nun who dared to prophesy the death of Henry VIII if he married Anne Boleyn; Oliver Cromwell (1661), who had passed away three years previously but was exhumed for a more public demise; and John Austin (1783), a highwayman, who was the last person hanged on the Tree.

THE SITE OF TYBURN TREE

The site of the Tyburn Tree at the junction of Oxford Street, Park Lane and Edgware Road

The Cock and Pye Ditch

The Cock and Pye Ditch, also known as the Marshland Ditch, was an ancient waterway that drained an area of St Giles called Marshland, which more or less corresponds with the picturesque junction now named Seven Dials. The ditch's precise origins are unknown, although it is referenced in a twelfth-century text describing the opening of a leper hospital at St Giles in the Fields on marshy land surrounded by ditches.

The ditch's unusual name is taken from a nearby pub, the Cock and Pye Inn, which supposedly served 'peacock in a pie', a luxurious dish that had the bird's head and tail feathers sticking out of its crust. Despite this exceptional local cuisine, Marshland was a filthy area and the ditch became a health hazard in the run-up to the Great Plague of 1665. It was covered over and converted into a sewer soon afterwards.

The site of the Cock and Pye Inn now accommodates modern restaurant.

Seven Dials

St Martin's Lane

Having collected all manner of detritus from the area around Seven Dials, the Cock and Pye Ditch ran down the length of Monmouth Street and then St Martin's Lane, before branching to the east and emptying into the Thames near Hungerford Bridge and Embankment station.

The ditch was the subject of some controversy a few years after it was converted into a sewer, when it transpired that a certain Richard Frith – a prominent builder after whom Frith Street is named – had illicitly connected the sewer from his housing developments in Soho with the St Martin's Lane section of the Cock and Pye Ditch. This unauthorised sewer was deemed to contribute far too much waste to the ditch and Frith was forced to start again.

The ditch passes beneath the church of St Martin-in-the-Fields, which has stood on the spot in various guises since at least the early thirteenth century, when it quite literally was in the fields between Westminster and the City of London. The sewer then flows parallel to Villiers Street and beneath the Victoria Embankment, which was developed in the 1860s in order to accommodate London's new sewage and transport systems. The incongruous stone steps at the southern end of nearby Buckingham Street would have led down to the water in the days when Dickens lived there.

The Cock and Pye Ditch was one of a number of sewage ditches that drained what is now the West End. On Carting Lane, there is a discrete reminder of the Southampton Sewer, which ran parallel to the ditch: a sewage lamp nicknamed Iron Lily, which was designed in the late Victorian period to burn off sewage smells from below.

Iron Lily on Carting Lane

Croydon Canal

4
Waterways of South London

1. Price's Candle Factory
2. The Falcon pub
3. Northcote Road
4. Tooting Bec Common
5. Streatham High Road
6. London Leather Exchange
7. Tower Bridge
8. St Saviour's Dock
9. Jacob's Island
10. Site of Bermondsey Abbey
11. Neckinger Mills
12. Manor House of Edward III
13. Southwark Park
14. Greenland Dock
15. Site of St Thomas à Watering pub
16. Old Kent Road
17. Millwall FC
18. Burgess Park
19. Peckham Square
20. Site of the Kentish Drovers pub
21. Asylum Road
22. Ruskin Park
23. Peckham Rye Park

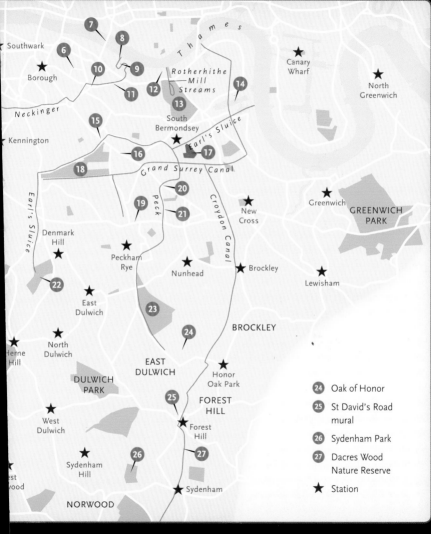

Thames

Southwark

7

8

6

10

9

Borough

11

12

Rotherhithe
Mill
Streams

14

Canary
Wharf

North
Greenwich

Neckinger

15

13

South
Bermondsey

Earl's Sluice

17

Kennington

16

18

Grand Surrey Canal

20

19

21

Peck

Croydon Canal

New
Cross

Greenwich

GREENWICH
PARK

Earl's Sluice

Denmark
Hill

Peckham
Rye

Nunhead

Brockley

Lewisham

22

East
Dulwich

23

24

BROCKLEY

Herne
Hill

North
Dulwich

DULWICH
PARK

EAST
DULWICH

Honor
Oak Park

West
Dulwich

25

FOREST
HILL

Forest
Hill

Sydenham
Hill

26

27

est
ood

Sydenham

NORWOOD

24	Oak of Honor
25	St David's Road mural
26	Sydenham Park
27	Dacres Wood Nature Reserve
★	Station

Falcon Brook

The Falcon is among London's longer streams, with sources as far south as Tooting and Streatham.

The southernmost source originates just south of Tooting Bec Common, in the vicinity of Thrale Road. It flows northwards along Dr Johnson Avenue and then follows the outline of the common and the railway track before crossing Balham High Road and making its way to Nightingale Lane via Mayford Road. This part of the stream was once known as the Woodbourne or Streatbourne, hence Streathbourne Road near the common.

A possible source of the Falcon at Streatham Hill and Downton Avenue

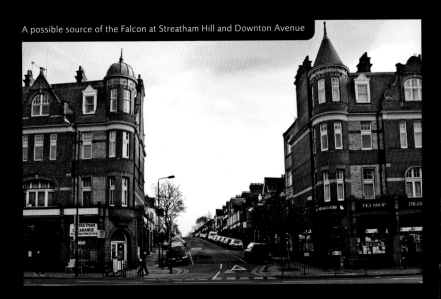

Looking into the valley of Northcote Road

The second stream – whose original name was Hydeburn or Hidaburna – starts at Streatham Hill, just north of Streatham Hill station, although local lore has it that there is a spring beneath a row of shops on Streatham High Road. Whatever the precise location of its source, this brook then flowed across residential areas of Streatham and Balham, via a remarkable number of roads whose names have watery connotations: Radbourne Road, Weir Road, Alderbrook Road and so on. It crossed Balham High Road just below Clapham South station and converged with the Tooting Bec branch on Nightingale Lane.

From here, the Falcon Brook followed the deep valley of Northcote Road and St John's Road all the way to what is now Clapham Junction station.

The Falcons of Clapham

The prevalence of Falcon-inspired streets and businesses in the
Clapham area is thought to have come from the St John family
– lords of Battersea Manor during the seventeenth century –
whose crest was a rising falcon.

The renaming of the Streatbourne and
Hydeburn streams as Falcon Brook seems to have occurred after
a riverside pub adopted the name Falcon Inn in the eighteenth
century. The pub was supposedly particularly popular with
local undertakers, so much hilarity ensued when a Mr Robert
Death took over as landlord in the early nineteenth century. An
engraving of 1801 by John Nixon depicts a crowd of drunken
undertakers outside a pub thought to be the Falcon Inn, with
the caption 'Undertakers regaling themselves at Death's Door'.
Patrons of today's Falcon pub are at considerably less risk
of falling into a river when they stumble out at closing time,
although the underground Falcon burst out of the pavement
in numerous places during the wet summer of 2007 and
temporarily turned Falcon Road back into a river.

Price's Candle Company

The Falcon at Clapham Junction

River Neckinger

The Neckinger appears to have risen near the Imperial War Museum, although the first half of its course is fairly unclear. It seems likely that the river ran along the route of the aptly named Brook Drive and then crossed the New Kent Road and Great Dover Street into Bermondsey.

In Bermondsey – which has been translated as 'island by a stream', perhaps due to its marshy surroundings – the Neckinger's legacy is more visible. Rapid industrial expansion in London around the turn of the nineteenth century saw much of the capital's leather workers collected together in this area south of the Thames, where the Neckinger would have provided access and running water. Tanner Street, Leathermarket Street and Leathermarket Court are among the roads whose names hark back to the time when Bermondsey was at the forefront of the British leather industry, and the formerly canalside Neckinger Mills building on Abbey Street was home to firms of leather specialists for almost 200 years.

The London Leather Exchange on Weston Street

Neckinger Mills in Abbey Street

Abbey Street's second claim to fame is that it is named after Bermondsey Abbey, a Benedictine monastery that was founded in 1082 on what is now the junction with Tower Bridge Road. The monks enlarged and embanked the swampy inlet at the Thames and built a watermill and a dock there, which they named after their patron saint, St Saviour. It is thanks to the monks that the Neckinger was ever navigable as far as the abbey. After the Dissolution of the Monasteries under Henry VIII, the mill supplied the local area with water. Later industrialisation saw it turned into the first water-powered gunpowder factory in England, and later replaced first by a paper mill and then by lead mills.

St Saviour's Dock and Jacob's Island

Although the Neckinger and its inlet were originally widened and maintained by Benedictine monks, the river's name has a rather less holy origin. In the eighteenth century, when piracy on the Thames was a common problem, condemned pirates were often hanged near the mouth of the river, an area popularly nicknamed Devol's Neckenger ('the devil's neckcloth') as a result. The corpses of the deceased were displayed further downstream as a deterrent to other would-be raiders.

The development and embankment of the area around St Saviour's Dock had created a man-made island known as Jacob's Island, which had become one of the most notorious slums in London by the nineteenth century. Populated by pickpockets and murderers and separated from surrounding areas by a network of putrid ditches – the largest of which, Folly Ditch, is now Mill Street – Jacob's Island particularly fascinated Charles Dickens, who gave Bill Sikes a memorable death there in *Oliver Twist* (1838). He described the slum as having 'dirt-besmeared walls and decaying foundations, every repulsive lineament of poverty, every loathsome indication of filth, rot, and garbage'. The housing estate that now stands on the site of Jacob's Island is called the Dickens Estate and its buildings are named after some of his most memorable characters.

By the mid-nineteenth century, Jacob's Island was so renowned for its sewage, dirt and cholera that most of the buildings were torn down and replaced by Victorian wharves, of which St Saviour's Wharf and New Concordia Wharf still survive.

St Saviour's Wharf

The Rotherhithe Mill Streams

The Rotherhithe Mill Streams began as a labyrinth of streams, ponds and small islands in the Southwark Park Road area. Near the bottom of West Lane they converged into a single stream, which flowed past the Manor House of Edward III on Bermondsey Wall East on its way to the Thames.

The Southwark Park area of Rotherhithe – formerly Redriff – was once called Seven Islands, although there were more than seven islands between the streams. Rather like Jacob's Island, they were largely man-made as a result of development along the watercourse, with sluices and mills dotting the final stretch down to the Thames. In the eighteenth

The remains of Edward III's manor house

A leisure centre is all that remains of the idyllic Seven Islands

century, Seven Islands was known as a tranquil haven for wildlife and became a popular spot for refreshments and rowing boats. Nearby tea gardens 'with music and dancing in the evening' proved a popular draw well into the eighteenth century, although they perhaps inevitably became better known for prostitution as time went on.

The stream crossed what is now Jamaica Road beneath Mill Pond Bridge, whose name is reflected in the nearby Millpond Estate. Just beyond here are the ruins of the Manor House of Edward III. This was built in the 1350s, although it was later replaced by a pottery factory and then a tobacco warehouse. Remains of the house were discovered during construction work and properly excavated and studied by English Heritage in 1985

Earl's Sluice

Earl's Sluice rises in Denmark Hill's Ruskin Park, which is named after the artist and critic John Ruskin, who lived nearby for much of his life. It flows across the Old Kent Road to meet the Thames near Deptford.

The stream's name evokes both nobility and sewage, which is apt considering its history. It was named after the twelfth-century nobleman Robert Fitzroy, first Earl of Gloucester and an illegitimate son of Henry I, who was Lord of the Manor of Camberwell and Peckham. It was converted into the Earl Main Sewer in the 1820s.

Ruskin Park

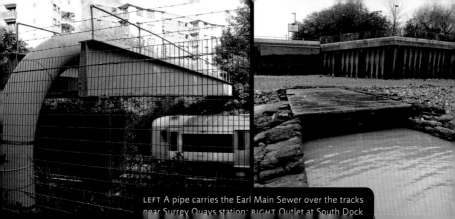

LEFT A pipe carries the Earl Main Sewer over the tracks near Surrey Quays station; RIGHT Outlet at South Dock

St Thomas à Watering

Earl's Sluice historically formed part of the boundary between the counties of Kent and Surrey, and so it is fitting that one of its most enduring contributions to British history should have occurred on the Old Kent Road.

In 1934, evidence of a medieval bridge was discovered in a trench at the junction of the Old Kent Road and Shorncliffe Road. This section of Earl's Sluice was nicknamed St Thomas à Watering after it became a popular horse-watering place for pilgrims on their way to the shrine of St Thomas à Becket in Canterbury. The most famous breather ever taken here was of course that of Chaucer's pilgrims in his late-fourteenth-century *Canterbury Tales*, who rode 'unto the watering of Seint Thomàs' and there decided who should tell the first tale. The Thomas à Becket pub stood on the site for many years but has recently been turned into an art gallery. A mural on the wall of a nearby civic centre depicts a scene from *The Canterbury Tales*.

Between the sixteenth and eighteenth centuries, this spot was a place of public execution. A notable execution here was that in 1539 of the Vicar of Wandsworth, his chaplain and two other members of his household, who were hanged, drawn and quartered for daring to question the supremacy of Henry VIII. Thomas Wyatt the Younger, who led Wyatt's Rebellion in 1554 and was hanged, drawn and quartered as a result, suffered a final indignity by having one of his quarters put on display here. The gallows at St Thomas à Watering was used for the last time in 1740 for the execution of two murderers, a father and son.

AND off WE RODE AT SLIGH-
TLY FASTER PACE than
WALKING TO ST. THOMAS'
WATERING-PLACE; AND
THERE OUR HOST DREW
UP, BEGAN TO EASE HIS
HORSE, AND SAID 'NOW
LISTEN IF YOU PLEASE;—

River Peck

The Peck, a tributary of Earl's Sluice (see p.108), rose between Honor Oak Hill and Peckham Rye and drained into the Sluice near South Bermondsey station.

The name Peckham is of Saxon descent and means 'the village of the River Peck'. The Peck itself was converted into a branch of the Earl Main Sewer along with Earl's Sluice in the 1820s, and the only remaining above-ground section of the river is in Peckham Rye Park. The poet and artist William Blake claimed to have seen a vision among the trees of Peckham Rye in 1767 when he was 10 years old, which he described as 'a tree filled with angels, bright angelic wings bespangling every bough like stars'. A tree that was christened the Angel Oak in his honour has long since died.

More recently, perhaps the most extraordinary sights in Peckham Rye Park were Italian prisoners of war, who were detained inside temporary huts here during the Second World War. One such hut still stands next to the café.

Further upstream, the source of the Peck is at the summit of Honor Oak Hill, which is named after an oak tree beside which Elizabeth I shared a picnic with a courtier in May 1602. The current Oak of Honor, which is surrounded by a railing, is a replacement for the original and was planted around the turn of the twentieth century. The summit of Honor Oak Hill was also used as part of a system of early-warning beacons during the reign of Elizabeth I, notably upon the arrival of the Spanish Armada in 1588.

The Peck in Peckham Rye Park

The Kentish Drovers

After following the gentle curve of the western side of
Peckham Rye Park, the Peck meanders northwards towards
its convergence with Earl's Sluice.

Just before reaching the Old Kent
Road, the Peck travels along most of Asylum Road, which is
named after an institution that was established there in 1827
under the patronage of the Duke of Sussex and later Prince
Albert. This asylum had a rather niche clientele: it was the
Licensed Victuallers' Asylum, in which retired victuallers who
had fallen on hard times, along with their wives or widows,
could see out their days in peaceful surroundings, paid for by
subscriptions from working members of the trade.

As a thriving town
on the main thoroughfare between Kent and London, Peckham
was a popular resting point for drovers taking their livestock
up to Smithfield Market (see p.36). The Old Kent Road was
lined with lively pubs, among them the Kentish Drovers, which
stood at the junction of the road and the Peck in the eighteenth
and nineteenth centuries, on the site of what is now the New
Saigon restaurant. Around the top of the building there is
still a colourful but weather-worn mural depicting a scene of
drovers passing along the grassy banks of the Old Kent Road,
surrounded by geese and other wildlife.

The Peck joins Earl's Sluice
shortly after passing the Kentish Drovers and there is little to
be seen of either between here and South Dock – although if
you listen carefully above the drains on Rope Street, you can
hear the streams on their way to the Thames.

The Kentish Drovers mural on Old Kent Road

ALICE DA

The Grand Surrey Canal

The Grand Surrey Canal was authorised by an Act of Parliament in 1801, at the height of a 20-year period labelled by historians as one of 'Canal Mania'.

In the early days of mass industry before the coming of the railway, the only conceivable way of moving goods from their places of manufacture to places where people could use or buy them was by water. A web of canals was hastily established across the nation, while in London the Thames's banks were broken at a number of places in order to link manufacturing towns with the great river and the wider world. In this context, it is perhaps unsurprising that the original plans for the Grand Surrey Canal saw it running all the way to Portsmouth.

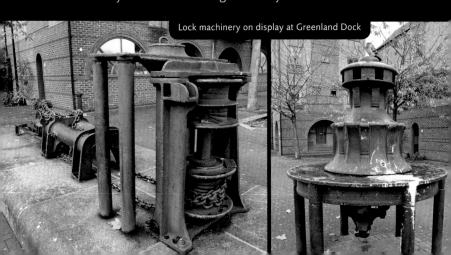

Lock machinery on display at Greenland Dock

In the end, the canal never even made it as far as Epsom, through which it was supposed to run on its way to the sea. The digging of the canal mouth at Rotherhithe was overshadowed after just a few years by grandiose plans to extend the various docks that ultimately made up the Surrey Commercial Docks (see p.174). The canal company saved time and money by incorporating the docks into their scheme, as well as by having the Croydon Canal (see p.120) branch off from the Grand Surrey Canal instead of making its own way to the Thames, but ultimately canals proved less of a financial draw than docks, and the whole thing was abandoned having got no further than Camberwell.

The Rotherhithe section of the Grand Surrey Canal, which extended as far as the Old Kent Road, opened in 1807. Although long since abandoned, there are numerous reminders of its former course along this route.

Burgess Park

The stretch of the Grand Surrey Canal that connected the Rotherhithe section and the Croydon Canal to Camberwell was opened in 1809, with a short extension to Peckham opened in 1826.

For the rest of the nineteenth century, the canal was a hub of the construction industry, with horse-drawn barges ferrying timber and limestone to the kilns and sawmills that lined its banks. There was further money to be made by imposing tolls, granting fishing licences and renting out rowing boats, while the towpath became a popular spot for illegal dog races. By 1811, the canal was so popular with illicit traders that 'bank rangers' were appointed to maintain order.

An 1816 lime kiln in Burgess Park

Grooves in the stonework of Hill Street Bridge in Peckham were made by horse-drawn barges on the canal

The Peckham branch ended in a basin that is now Peckham Square, while the Camberwell branch terminated at Camberwell Basin. The latter covered much of what is now Burgess Park, which is still home to an iron footbridge that once spanned the canal. A large factor in the park's development was the decline of the Grand Surrey Canal during the Second World War, when neighbouring houses and industrial buildings sustained heavy bombing damage. The final stretches of open water were drained in the 1960s and 1970s after they became infamous dumping grounds for rubbish and the occasional dead body.

The Grand Surrey Canal is one of the most rewarding of London's lost waterways for urban explorers hunting for relics. Bridges and timber merchants line its route – including a former wharf still used by timber merchants behind Peckham Square – and there is even a sizeable mooring bollard on Surrey Canal Road in New Cross.

Croydon Canal

The plans for Croydon Canal were given the go-ahead in 1801, at the same time as the Grand Surrey Canal (see p.116). It was originally supposed to start at the Thames in Rotherhithe but financial concerns persuaded the developers to branch it off from the Grand Surrey Canal in New Cross. Croydon Canal opened in 1809 and culminated in a basin on the site of what is now West Croydon station.

The main use of the canal was for the transportation of lime, timber, chalk and agricultural produce into London. The canal was an overly ambitious project, however, and the 28 locks it required between New Cross and Honor Oak proved incredibly expensive to run. They also created the nineteenth-century equivalent of traffic tailbacks, with barges queuing to

The junction of Croydon Canal (right) and the Grand Surrey Canal in New Cross

A mural in David's Road, Forest Hill

proceed over this two-mile course. Shares in the canal, which had started out at £100 each, had fallen to the equivalent of 10 pence by 1830. Croydon Canal was destined to be a financial failure, and it closed in 1836.

In New Cross, the routes of the two canals are best appreciated at the junction of Mercury Way (Croydon Canal) and Surrey Canal Road (Grand Surrey Canal). The two canals were connected with Rotherhithe via a sharp bend at what is now the junction with Canal Approach.

In Forest Hill, a recent mural alongside David's Road commemorates the canal's former route along that road. The nearby Dartmouth Arms pub – built in the 1860s to replace an 1815 original – was where the body of Mary Clarke, suspected of having been 'foully murdered' by the father of her unborn child, was formally identified after it was discovered floating in Croydon Canal in June 1831.

Down to Croydon

The abandoned Croydon Canal was turned into a railway line, which opened in 1839. The track that connects Surrey Quays and West Croydon stations more or less follows the canal's route, a common practice during the widespread conversion of old canals since they tended to be straight and the right sort of width, with conveniently level beds. There were some short lengths of Croydon Canal that were unsuitable for incorporation into the railway, and these survived a little longer as pleasure-boating lakes before being buried by the growing suburbs.

The canal had two reservoirs on this lower stretch – one at Sydenham and one in South Norwood – which ensured an even supply of water along the whole route. The Sydenham reservoir occupied much of the site of what is now Sydenham Park and was a popular place for swimming and ice-skating. The South Norwood reservoir still exists as South Norwood Lake.

There are further traces of Croydon Canal in Dacres Wood Nature Reserve between Forest Hill and Sydenham, where a kink in the canal once formed the railwayside gardens of a Victorian house named Irongates. Betts Park in Anerley incorrectly signposts a short fragment of the canal – now encased in concrete – as the last visible section.

There is very little to be seen of the canal as it approaches its basin at West Croydon station, although the names of Towpath Way and Canal Walk, now alongside the railway line, indicate the route it took.

Betts Park

The Black Ditch

5
Waterways of East London

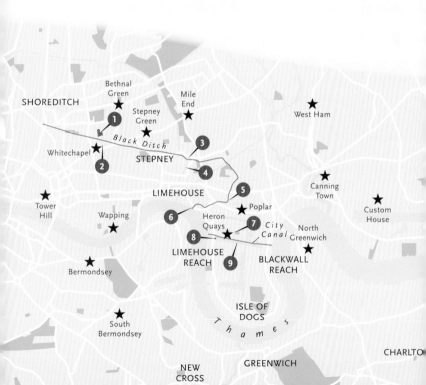

1. Jewish Cemetery
2. Ducking Pond Row
3. Rhodeswell Road
4. Regent's Canal
5. Site of Stonebridge Pond
6. Limekiln Dock
7. Canary Wharf
8. The City Pride pub
9. South Dock

SHOREDITCH

Bethnal Green

Mile End

West Ham

Stepney Green

1 Black Ditch

Whitechapel

STEPNEY

2

3

4

LIMEHOUSE

5

Canning Town

Tower Hill

Wapping

6

Heron Quays

Poplar

7 City Canal

North Greenwich

Custom House

8

9

Bermondsey

LIMEHOUSE REACH

BLACKWALL REACH

South Bermondsey

ISLE OF DOGS

Thames

NEW CROSS

GREENWICH

CHARLTO

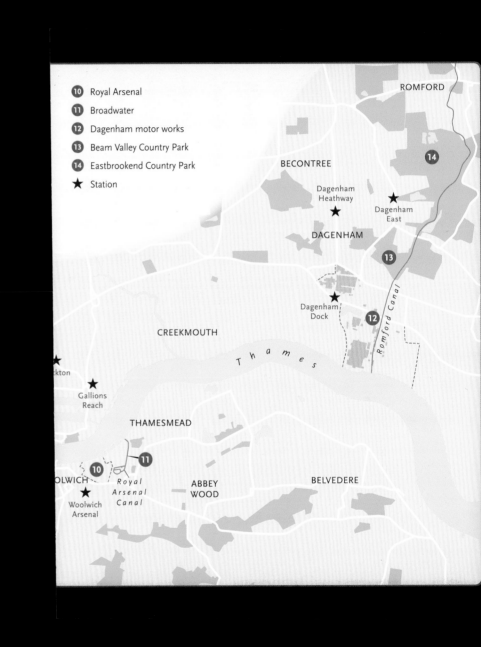

10 Royal Arsenal
11 Broadwater
12 Dagenham motor works
13 Beam Valley Country Park
14 Eastbrookend Country Park
★ Station

The Black Ditch

The Black Ditch is known for certain to have formed a crescent shape between Stepney and Limehouse, but it is possible that its source was as far back as Holywell Lane in Shoreditch, which was also the source of the Walbrook (see p.42).

Evidence for this western section is fairly scant – Daniel Defoe mentioned 'a piece of ground just over the Black Ditch' in the early eighteenth century and a map of 1799 appears to indicate it – but it would have crossed over what is now the junction of Whitechapel Road, Cambridge Road and Mile End Road on its way to Stepney. At Brady Street, now adjacent to a Jewish cemetery, water from

The Jewish Cemetery near the ducking pond

The Black Ditch crossing Regent's
Canal at Rhodeswell Road

the Black Ditch is thought to have pooled into a ducking pond,
where petty criminals were strapped to a chair and dunked
as a form of punishment. Nearby Durward Street appears as
Ducking Pond Row on a map at the turn of the nineteenth
century, although it was subsequently renamed Buck's Row
and then renamed again in 1888 after gaining infamy as the site
of Jack the Ripper's first murder.

        The majority of maps show the Black Ditch
originating in Stepney at a pond on Rhodeswell Road, which was
once named Rogue's Well. 'Stepney' itself is supposedly derived
from 'Stybbanhythe', after a Saxon warrior called Stibba who
navigated up what is thought to have been the Black Ditch in the
tenth century and established a settlement there.

Down by the Docks

The Black Ditch continued east before looping back on itself and meeting the Thames at Dunbar Wharf. A stone bridge over the ditch at the western end of Poplar High Street was documented in the mid-fifteenth century, and it was still a watering place four centuries later, when an inn called the White Horse stood beside a Stonebridge Pond. The last incarnation of this pub closed down in 2003, but its sign still stands outside a block of flats.

The Black Ditch emptied into an inlet called Limekiln Dock, and the name Limehouse comes from the lime oasts that produced quicklime here in the fourteenth century. Thanks to its riverside location, Limehouse was historically a focal point for both import and immigration. During the nineteenth century, an increase in trade with China led to the establishment of London's first Chinese community here,

The White Horse in Poplar

Limekiln Dock

City Canal

City Canal, sometimes referred to as the Isle of Dogs Canal,
once cut across the top of the Isle of Dogs to create a shortcut
between Blackwall Reach and Limehouse Reach.
The canal was
commissioned in 1799 and completed in 1805, its purpose
being to appease merchants who complained that London's
wharves and warehouses were too far inland. The three-mile
loop past Greenwich not only lengthened every journey into
and out of London quite considerably, but it also proved
difficult to negotiate when the tide was going in an unhelpful
direction. The threat of lucrative trading contracts being
relocated to more accessible cities was enough to convince
the government to authorise this ambitious scheme.

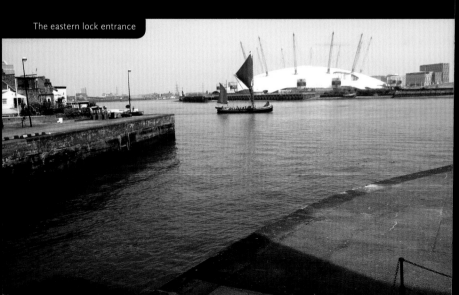

The eastern lock entrance

The view west across what was once City Canal

The City of London Corporation bought the land and constructed the canal, sweeping aside numerous buildings associated with the area's fishing industry in order to make way for dry docks, shipyards, timber yards, iron foundries and a lock at either end of the waterway. To this day, the eastern lock entrance at Blackwall Reach is the only working entrance from the Thames into the Isle of Dogs docks, although the current blue swing bridge is the sixth bridge to stand on that spot

From Canal to Dock

Although City Canal looked like a time-saving shortcut on paper, the realities of the tidal Thames meant that using the canal could take almost as much time as going the long way round. Like so many of the waterways constructed during the period of 'Canal Mania' (see p.116), City Canal soon became a financial disaster, unable to attract enough business to pay for its own upkeep. In 1829 it was sold to the company that owned the neighbouring West India Docks (see p.184), whose main motivation was to prevent the canal from falling into the hands of a rival dock company.

The West India Dock Company closed the canal, enlarged the watercourse and linked it with the existing dockyard to create the South West India Dock (nowadays South Dock). The line of South Dock more or less follows the canal's original line, although the southernmost buildings by Heron Quays station are built directly over what was once City Canal.

In 1931 the eastern lock entrance was rebuilt and the various docks of the Isle of Dogs were linked together. Around the same time, the western entrance was formally blocked off by the construction of a road and a pumping station that controlled the level of the water within the docks. In the mid-1930s, the City Arms pub that had been built near the western entrance during the canal's brief heyday was rebuilt and renamed the City Pride, which it remains to this day.

At the far side of the pumping station, the western entrance at Limehouse Dock nowadays serves as an impounding dock.

Water-depth markers carved into
the canal wall at Limehouse Dock

The Royal Arsenal Canal

The Royal Arsenal Canal was a short waterway connecting the Royal Arsenal at Woolwich with the Thames. Due to its main use for the transportation of military supplies, it was also known as the Ordnance Canal, while a third name, the Pilkington Canal, is a tribute to the colonel who oversaw its development. The remaining section of the canal that cuts into the 1960s development of Thamesmead is now known as Broadwater.

Woolwich Warren – a storage depot for artillery supplies – was established in the 1670s, and renamed the Royal Arsenal on the suggestion of George III in 1805. Convict labour had been used to construct a 2.5-mile boundary wall at the end of the eighteenth century, and it was used again to dig the canal during the 1810s. These convicts were housed on prison hulks.

Most of the arsenal's gunpowder was supplied by the Royal Gunpowder Mills (see p.146) on vessels that arrived via the Lea and the Thames and accounted for much of the traffic on the canal. In addition, military ships would collect firepower and other armaments on their way out to the naval shipyard at Chatham, which fitted weaponry onto military vessels.

Since the canal's closure in the 1960s, Broadwater has been a purely ornamental feature, albeit one that attracts more than its fair share of rubbish and abandoned furniture. The former entrance from the Thames has been blocked over by a concrete wall but there are still various remnants to be seen, including the entrance lock, a swing bridge and mooring bollards.

Broadwater

A derelict swing bridge by the canal entrance

Romford Canal

The other canals in this book reveal stories of sad decline, but Romford Canal is unique among the forgotten canals of London in that it never officially opened.

The allure of Romford was that it was one of the main potato-growing areas for the London market, and the canal was going to make the transportation of crops into London and manure out to the farmers much more efficient. At a time when the widespread conversion of canals into railways was underway, authorisation was granted for this (in hindsight) foolhardy scheme.

Beam Valley Country Park

The overgrown course of the canal at Eastbrookend Country Park

The canal's course was meant to run northwards from the Thames near the modern-day Dagenham motor works to Romford in Essex, closely following the route of the Rom (or Beam) river. The first plans were drawn up in 1809 and construction eventually began in 1875, only to be abandoned two years later when funds ran out. Of the original plans envisaging a busy waterway wide enough to be navigated by 60-ton barges, only the lower section had been completed by the time the project was shelved, but this included two locks, two bridges and a tunnel.

Nowadays, there is no longer a great deal to see of the ill-fated canal. There is a short section containing some water at the Beam Valley Country Park in Dagenham, and a dry section of the canal bed can be seen at Eastbrookend Country Park near Elm Park.

⚠ **Warning**
Confined space

🚫 No unauthorised
entry

River Moselle

6
Tributaries of the River Lea

Palmers Green ★

Pymmes Brook

Angel Road ★

See inset map right for route north of Banbury Reservoir

Arnos Grove ★

Bounds Green ★

Muswell Stream

Lesser Moselle

White Hart Lane ★

6

7

Carbuncle Ditch

BANBURY RESERVOIR

MUSWELL HILL

1

Wood Green ★

Moselle

ALEXANDRA PARK

2

3

Highgate ★

5

4

Moselle

8

WALTHAMSTOW

9

Seven Sisters ★

St James Street ★

Crouch Hill ★

FINSBURY PARK ★

Stamford Hill ★

Lea

11

12

13

Arsenal

STOKE NEWINGTON

Clapton ★

14

CLAPTON

10

Tufnell Park ★

HACKNEY

Hackney Brook

Hackney Wick ★

Hackney Central ★

15

16

17

18

Pud Mill ★

Pudding Mill River

Highbury & Islington ★

Camden Town ★

King's Cross ★

Hoxton ★

Bethnal Green ★

BOW

Bow Road ★

Bromley-by-Bow ★

Old Street ★

Stepney Green ★

1. Muswell Road
2. Queen's Wood
3. Priory Park
4. Lordship Recreation Ground
5. Broadwater Farm Estate
6. Tottenham Cemetery
7. White Hart Lane stadium
8. Tottenham Marshes
9. Markfield Recreation Ground
10. Old Arsenal stadium
11. Clissold Park
12. Abney Park Cemetery
13. Stamford Hill
14. Hackney Downs
15. Mare Street
16. Victoria Park
17. Old Ford Lock
18. Olympic stadium
19. Royal Gunpowder Mills
★ Station

The Royal Gunpowder Mills Canals

The Royal Gunpowder Mills Canals were a system of waterways that served the extensive ammunitions complex of the same name at Waltham Abbey, just outside the M25.

This canal system was developed out of the various streams that ran alongside and flowed into the River Lea, which itself provided vital access into and out of central London. Millhead Stream was one of the existing watercourses that were incorporated into the canals. It had been used for transportation from the earliest days of the mills (see overleaf), and in 1806 it was connected to numerous new buildings

A lock leading up to the drained high-level canal and a 1906 building that replaced one destroyed in a 'terrible explosion'

A footbridge spanning a drained canal

within the complex by newly dug canals – including one named
the Powdermill Cut – which made the transfer of materials both
between buildings and towards the Lea much more efficient.
A survey undertaken in 1814 counted 'five barges, nine powder
boats, two ballast barges and six punts' on the site.

The overall length of
the canal system in the nineteenth century was just over five
miles. High- and low-level waterways supported by a system of
aqueducts and locks kept boats carrying gunpowder separate from
ones carrying raw materials. Safety was obviously paramount, and
the use of canals rather than bumpy roads to transport gunpowder
was as much a precaution against catastrophic bumps and crashes
as it was a time-saving scheme. The barges were even towed by
men, to minimise the risk of a bolting horse causing an explosion.

The Gunpowder Mills

The Gunpowder Mills began life as a cloth-production mill
run by the monks of Waltham Abbey in the sixteenth century.
It was converted into an oil mill in the early seventeenth
century, but the increasingly belligerent nature of international
relations meant that gunpowder production became more of
a priority as that century progressed.

Between the late eighteenth
and late nineteenth centuries, the complex grew along the
banks of Millhead Stream in an early version of an assembly
line. Its efficiency was remarkable for the period, and it was
described in 1735 as 'the largest and compleatest works in
Great Britain'.

In 1787, the Gunpowder Mills were sold to the Crown,
which retained ownership over the site until it was closed down
in 1991. This purchase was the idea of the Deputy Comptroller
of the gunpowder laboratories at the Royal Arsenal in Woolwich
(see p.136), and it paid off because the Royal Gunpowder Mills
went on to become the arsenal's major supplier of gunpowder.
The mills produced munitions for Waterloo, the Crimea and
the Boer Wars, and also produced explosives for quarrying
and tunnelling purposes.

The First and Second World Wars were
times of great activity for the Royal Gunpowder Mills, and the
increase in industry and employees – over 6,000 during the
First World War, most of them women – also necessitated the
extension of the canal system to a length of around 10 miles.
In 1945, however, the facility was turned into a research and
development centre for rockets and explosives, which it
remained until 1991.

The bridges were designed to accommodate the powder boats' barrel roofs

Powder boats were towable from either end, which avoided having to turn them

Muswell Stream

As its name suggests, Muswell Stream rises in Muswell Hill.
It then crosses Bounds Green and the North Circular before
emptying into Pymmes Brook, a tributary of the Lea, behind
the bus garage in Palmers Green.

Muswell Hill is thought to be named
after the hill on which a 'mossy well' once stood. The well itself
supposedly had healing qualities, and the land surrounding
it was given to an order of nuns in the twelfth century by the
Bishop of London. They built a chapel there called Our Lady of
Muswell and used the land for dairy farming.

The well at Muswell was
the main source for the three headstreams that converged to
form Muswell Stream. Nowadays, the capped well is located

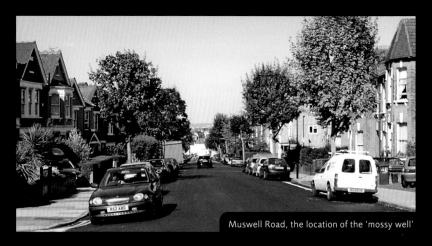

Muswell Road, the location of the 'mossy well'

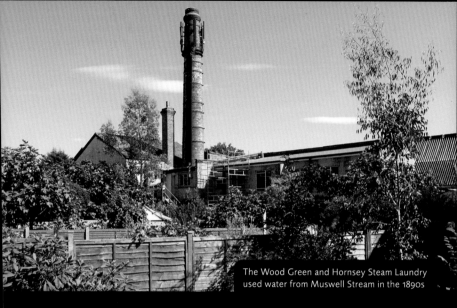
The Wood Green and Hornsey Steam Laundry used water from Muswell Stream in the 1890s

beneath a house on Muswell Road. One of these three streams was channelled into a series of ornamental lakes in Alexandra Park in 1875, although they were drained in order to make way for Grove Avenue and Rosebery Road around the turn of the twentieth century.

The lower reaches of Muswell Stream were covered over in the 1920s and 1930s, although this has not prevented the whole area from flooding during heavy storms. One suggested remedy was to widen the pipes carrying the stream beneath North London, but this would simply have exacerbated the flooding problem downstream. In the end, large water tanks were constructed beneath Woodside Park and Albert Road, which collect a vast volume of water over a short period of time and then release it slowly along the pipes.

River Moselle

Among the longest of the Lea's lost tributaries, the Moselle
twists and turns its way from Highgate to South Tottenham.
London's lost Moselle has nothing to do with the better-known
river of the same name in France: like Muswell Stream (see
p.148), the Moselle takes its name from the mossy well and hill
that spawned Muswell Hill.

Various streams from the hills of Highgate
and Archway converge to form the Moselle at Priory Park in
Hornsey. These streams include Cholmeley Brook, whose
sources are just north of Waterlow Park and at the street aptly

One of the Moselle's headstreams in Queen's Wood

A remainder of a bridge across the Moselle on Vincent Road

named Hillcrest; Priory Brook, which rises by Highgate station; and Etheldene Stream, which comes from the Queen's Wood area. The names of all three streams are reflected in numerous nearby streets.

That some of the Moselle's headstreams are still visible is indicative of the fact that the river is not entirely lost, and indeed the middle of its course is partially above-ground. Over three-quarters of the river is culverted, however, and the visible sections, as we will see overleaf, are very far from resembling the fish-filled 'sparkling Moselle' of yore.

After crossing Hornsey High Street just north of Priory Park, the Moselle makes its way towards Wood Green station, where it branches off and runs between Moselle Avenue and Lordship Lane.

The Deteriorating Moselle

After joining Lordship Lane at the end of Moselle Avenue, the Moselle flows alongside Downhills Way before cutting across Lordship Recreation Ground, where the deterioration of the river is at its most visible.

On either side of the Rec, the Moselle passes through a grille as it exits and then re-enters its culverted sections. While the grilles are clogged with shopping trolleys and fallen branches, the water itself is largely made up of an unpleasant-smelling green sludge.

LEFT The Moselle passing into a culvert beneath Broadwater Farm Estate; RIGHT Lordship Recreation Ground

LEFT The Moselle near Tottenham Cemetery; RIGHT Tadpoles ignore the swimming ban in the Lesser Moselle's lake

Between the Rec and Tottenham Cemetery, the Moselle passes beneath Broadwater Farm Estate, the former farmland that was transformed into a 12-block housing estate in the late 1960s. The area had always been deemed unsuitable for construction due to flooding from the Moselle, and the buildings of the estate had to be built on stilts despite extensive culverting along this stretch of the river. The estate had fallen into considerable decay by the mid-1970s but it achieved even greater notoriety in October 1985, when residents inspired by the previous week's Brixton riots mounted an ultimately violent uprising against the police, which resulted in the infamous murder of PC Keith Blakelock.

As the Moselle approaches Tottenham Cemetery, it receives a tributary known as the Lesser Moselle, which rises in the playing fields north of Lordship Lane. The Lesser Moselle itself is fed by the overflow from the cemetery's ornamental lake, which is also falling into poor condition.

Tottenham Cemetery and White Hart Lane

The most attractive stretch of the Moselle bisects Tottenham
Cemetery, at either side of which the river is once again
culverted. Tottenham Cemetery is located on the site of a well
that was famed for its healing waters until the nineteenth century,
when, along with most of London's other rivers, the Moselle
became too polluted for household use.

The Moselle runs eastwards
beneath White Hart Lane station and then branches southwards
down Tottenham High Road. In 1836, this stretch around White
Hart Lane Stadium was covered over, with much of the rest of the
river being buried by the end of that century. Polluted water and
the risk of flooding were the motives behind this scheme, but the
water made a habit of bursting its pipes and flooding Tottenham
High Road until the late 1960s, when the culvert was renovated.

Tottenham Cemetery

A footpath between the skate park and the Moselle in Markfield Recreation Ground

Below the junction with Lansdowne Road, the Moselle partially drains into Pymmes Brook via Carbuncle Ditch (see p.156). There is a final glimpse of the river further south behind Markfield Recreation Ground's skate park, which was built on the derelict remains of a sewage works.

Unlike the majority of lost rivers in this book, the Moselle's main use is as a storm drain for rain run-off from gardens and roads rather than for sewage. There have, however, been reports of incorrectly plumbed houses releasing all of their waste water into the river, which has a knock-on effect on pollution in Pymmes Brook, the Lea and the Thames.

Carbuncle Ditch

This man-made ditch, variously named Carbuncle Ditch and Garbell Ditch, was dug in the fifteenth century to ease the problem of flooding along the course of the Moselle (see p.150).

Carbuncle Ditch had settled on its current name by the nineteenth century, although records as to why such an unusual name was chosen seem not to exist. It possibly stems from the fact that the ditch was an unsightly growth on the side of the Moselle, with both watercourses becoming increasingly polluted in the days before sanitary sewerage.

LEFT Carbuncle Passage; RIGHT Carbuncle Ditch

The grassy path across Tottenham Marshes follows the course of Carbuncle Ditch

The ditch branches off from the Moselle at Scotland Green, just a few blocks south of White Hart Lane Stadium. It runs in an almost-straight line from here to Pymmes Brook, a tributary of the Lea, via Carbuncle Passage and across Tottenham Marshes. Carbuncle Passage, a wide walkway running along the backs of two rows of houses, was constructed in the early twentieth century over the course of the ditch. Drain inspection covers along the passage confirm the ditch's route.

There is a brief glimpse of the unculverted ditch between the end of Carbuncle Passage and the railway line running from Tottenham Hale to Northumberland Park, and the final view (and smell) of it is at its overgrown outfall at the far edge of Tottenham Marshes.

Hackney Brook

Hackney Brook was a waterway equal in length to London's largest lost rivers, which flowed across Stoke Newington and Hackney before meeting the Lea.

The brook, which was covered over and mostly turned into a sewer in the mid-nineteenth century, has two sources just off Hornsey Road in Holloway: the first at the Wray Crescent end of Tollington Park and the second near the junction of Holloway Road and Jackson Road. The two sources converge just west of Arsenal station and then cross Gillespie Road in front of Arsenal's old stadium before following Riversdale Road up to Clissold Park.

The two lakes at the north of Clissold Park indicate Hackney Brook's former route, although these water features are nowadays fed by mains water rather than by the brook itself. The park was originally the garden of Paradise House, the home of a Quaker and anti-slavery campaigner named Jonathan Hoare in the 1790s, and later of a Reverend Clissold, who renamed the entire estate after himself. Hackney Brook was already a polluted nuisance in Hoare's day, and he diverted the water's course away from his house – now called Clissold House – which still exists as the premises of Clissold Park Café.

With Hackney Brook now redirected underground, there is little to remind the residents of Stoke Newington that a thirty-foot-wide stream once flowed through the neighbourhood, although road names such as Grazebrook Street provide hints as to its route.

Clissold Park

Abney Park Cemetery

The northern perimeter of Abney Park Cemetery in Stoke Newington is defined by the course of Hackney Brook. Alongside Kensal Green (see p.72) and West Norwood (see p.52), both of which also stand on the banks of lost rivers, Abney Park is one of the 'Magnificent Seven' cemeteries of London.

Abney Park was once the grounds of Abney House, the second home of Lady Mary Abney and her husband, Sir Thomas, who was Lord Mayor of London at the turn of the eighteenth century. Lady Abney is credited with the transformation of the parkland into landscaped gardens and walkways, a task in which she was helped by family friend Sir Isaac Watts, a nonconformist and hymnwriter who authored, among many others, 'Joy to the World' and 'O God, Our Help in Ages Past'.

Abney Park Cemetery, which was established in 1840, has become known as a genuinely wild arboretum, with overgrown paths meandering among the graves and statues, but it was something of a nature reserve even in Lady Abney's day. Sir Isaac Watts was supposedly particularly taken by a heronry on an island in Hackney Brook at the northern end of the park, and took his inspiration from the beautiful scenery there.

Stamford Hill, on the eastern side of the cemetery, is, like Stamford Brook (see p.66) and Stamford Bridge (see p.73), named after a 'sandy ford' across Hackney Brook. On the far side of this former crossing, the brook runs along the northern side of Stoke Newington Common and then enters Hackney via the western perimeter of Hackney Downs.

This path marks the course of Hackney Brook along Hackney Downs

Abney Park Cemetery

Haca's Island

The name Hackney conceals a reference to the lost brook, *ey* being an ancient word for 'island'. 'Hack' seems to come from a Dane named Haca who once owned the land in this area. So 'Haca's Island' or even 'Haca's Marsh' are reasonable translations of the word.

Hackney Brook crosses Mare Street approximately where Hackney Central station stands today. This street is, perhaps surprisingly, not named after a horse but after a pond (or 'mere') that was fed by the brook and used for ducking petty criminals during the seventeenth century.

Hackney Brook flows from left to right across this junction at Mare Street

Hackney Brook's outfall near Old Ford Lock, with the sewage pipes from Mare Street passing overhead

There is a mid-fifteenth-century reference to it as Merestret. Although this junction is fairly unremarkable these days, the picturesque landscape through which Hackney Brook once passed can be seen in an old illustration in front of the sixteenth-century St Augustine's Tower.

Just beyond here, Hackney Brook was joined by Pigwell Brook, whose source was a well on Well Street. Numerous other roads in this area – Shore Place, Brook Street and so on – refer to the route of Hackney Brook.

The sewer that was constructed out of most of Hackney Brook's course branches off at Mare Street and crosses Victoria Park to meet the larger sewer network in Bow. The last stretch of the brook meanders north of the park and through Hackney Wick before joining the Lea by Old Ford Lock.

Pudding Mill River

Pudding Mill River was part of the Bow Back River network just north of Bow. It ran the short distance from Old Ford Lock to St Thomas's Creek at modern-day Stratford High Street and was a minor tributary of the Lea.

The pudding mill that gave the river its name was a small riverside complex of wind-driven corn and malt mills, whose name in turn was taken from the supposedly pudding-shaped thirteenth-century St Thomas's Mill. All of these mills have since been demolished, the majority of them in the early nineteenth century. A nearby windmill called Nobshill Mill was demolished in the 1890s, while the junction with St Thomas's Creek was closed off as part of the River Lea Flood Relief Scheme in the 1930s.

The Pudding Mill River gave its name to Pudding Mill Lane and (much later) a Docklands Light Railway station, although the river itself became increasingly unnavigable between the 1960s and the 1980s, by which time it was an overgrown rubbish-filled ditch beside a nuclear reactor.

The Olympic Stadium in 2010

At the turn of the twentieth century, regeneration work along Pudding Mill River saw trees planted and the remaining sections of open water becoming home to waterfowl and fish. London's successful bid to host the 2012 Olympic Games truncated this redevelopment, however, and from 2007 onwards the Pudding Mill River area became a building site. The wildlife was relocated to the nearby Lea and the Olympic Stadium was built on top of the infilled river.

All that remains to be seen is a small stub of the river where it meets the Lea.

7
Docks and Wharves

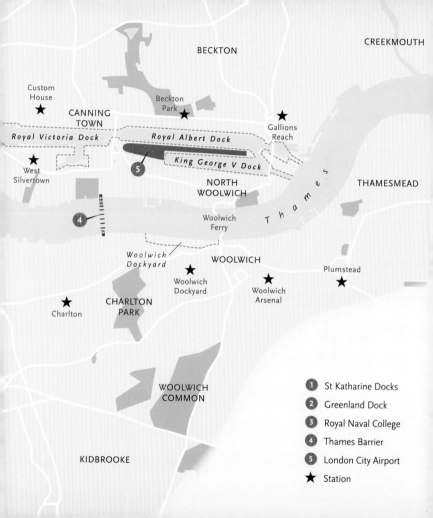

BECKTON

CREEKMOUTH

Custom House

★

CANNING TOWN

Beckton Park

★

Gallions Reach

★

Royal Victoria Dock

Royal Albert Dock

King George V Dock

5

West Silvertown

★

NORTH WOOLWICH

T h a m e s

THAMESMEAD

4

Woolwich Ferry

Woolwich Dockyard

WOOLWICH

Plumstead

★

Woolwich Dockyard

★

Woolwich Arsenal

★

Charlton

★

CHARLTON PARK

WOOLWICH COMMON

KIDBROOKE

1 St Katharine Docks

2 Greenland Dock

3 Royal Naval College

4 Thames Barrier

5 London City Airport

★ Station

The London Docks

The London Docks in Wapping, a vast complex consisting of six interlinked basins and docks, were constructed in the early nineteenth century. Until the neighbouring St Katharine Docks opened in 1828, they were the closest docks to the centre of London.

The first part of the docks to open was the 20-acre Western Dock in 1805, which had access to the Thames via Wapping Basin, followed by Tobacco Dock in 1813. Hermitage Basin opened in 1821 and provided a second entrance to the Thames. Eastern Dock, which was linked to Western Dock by Tobacco Dock, was added in 1828, and Shadwell Basin gave the complex a third entrance point to the Thames in 1832. The enormous amount of earth that was removed in order to dig the docks was spread onto the former marshland that is now Battersea Park.

Tobacco Dock

A view from Shadwell Basin towards the Thames

The Port of London

All of London's docks and wharves are part of the enormous Port of London, which was the world's largest port during the nineteenth century. At the height of its prominence, the Port of London effectively covered both banks of the Thames all the way from central London to the open sea.

Shipping had been vital to London's economy since Roman times, but a massive increase in international trade during the eighteenth and nineteenth centuries meant that the city desperately needed to accommodate more vessels. Whereas sailing ships had once been able to berth alongside wharves on the Thames, the numerous new docks that were constructed between Wapping and Woolwich around the turn of the nineteenth century provided a more secure environment in

Replicas of eighteenth-century ships *Three Sisters* and *Sea Lark* at Tobacco Dock

which to load and unload goods, and also helped decongest the Thames.

The Port of London was large enough to cope well with the transition from sailing vessels to steam-powered ships in the late nineteenth century. By the early twentieth century, however, cargo ships had become so enormous that the capacity of the London Docks had dwindled to 30 vessels. When container ships arrived, the docks were no longer a realistic option, being too small and too far from the sea.

Much of the London Docks was infilled in the late 1960s. Wapping Basin is now a sports pitch and Eastern Dock is a park. News International's extensive print works is on the site of Western Dock. A canal runs the short distance from Hermitage Basin to Tobacco Dock, both of which are now ornamental water features.

The Surrey Commercial Docks

Before Rotherhithe became the residential area it is today, its marshy ground was good for very little except docking ships. The Surrey Commercial Docks, which were begun at the end of the seventeenth century, gradually turned most of the land in Rotherhithe into a series of large interlinked pools of water.

The first of the docks was dug in 1696 and named the Howland Great Wet Dock in honour of the family that owned the land. The 10-acre dock had a capacity of 120 vessels and was originally intended for berthing and refitting ships rather than unloading cargo. The South Sea Company leased the Great Wet Dock in 1725 and used it as a base for its whaling ventures off Greenland, which were ultimately unprofitable but the dock to this day retains the name Greenland Dock.

After a further nine docks had been squeezed in, many of them also took on names inspired by the territories in which their owners operated: Russia Dock, Canada Dock, Quebec Dock, Norway Dock and so on. Greenland Dock briefly became Baltic Dock in 1807 when it was appropriated for incoming timber from that part of the world.

As part of this grandiose docking development, the Grand Surrey Canal (see p.116) was planned and partially constructed in the early years of the nineteenth century. It succeeded in converting Rotherhithe into a bustling network of interlinked docks and waterways but was a failure further south.

Russia Dock's mooring chains, bollards and depth gauges are still visible at Russia Dock Woodland

Rotherhithe Emerges

The process of converting Rotherhithe back into land began after the Second World War, during which London's Docklands suffered very heavy bombing. The cost of rebuilding hundreds of sheds and warehouses across the whole Port of London was hefty enough, but adapting all of the docks for the handling of new container ships was simply impossible, and the Surrey Commercial Docks – along with the others in this chapter – fell into disuse and decline during the 1960s.

Rotherhithe remained largely derelict throughout the 1970s but was saved in 1981 when the government established the London Docklands Development Corporation (LDDC) and charged it with putting the Docklands

The Dock Manager's Office

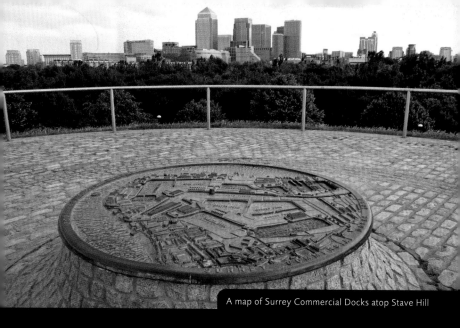
A map of Surrey Commercial Docks atop Stave Hill

to good use. By the time the LDDC had withdrawn from the area in 1998, the Docklands had been transformed into the hub of business, shopping and industry that it is today.

Under the LDDC, most of Rotherhithe's docks were filled in and the land used for residential blocks. Russia Dock became parkland and Stave Dock was turned into an artificial hill built of waste material and rubble. The late-nineteenth-century Dock Manager's Office on Deal Porters Way was also restored. The last remaining sections of water in what is now called Surrey Quays include Canada Water (Canada Dock), a freshwater nature reserve; South Dock (East Country Dock), London's largest marina; and Greenland Dock, a watersports centre.

Millwall Dock

The horizontal Outer Dock and vertical Inner Dock of Millwall Dock form an inverted L-shape in the middle of the Isle of Dogs. Work began in 1865 and was completed in 1868, with the mud and silt that was dug up being conveyed to a nearby area still known as Mudchute.

Millwall Dock originally had a capacity of almost 36 acres of water and was mainly used by grain traders, and indeed 'Millwall' is a reference to the windmills built alongside the Thames in the nineteenth century. Flour mills, granaries and warehouses lined the banks of the dock, although the last vestiges of this trade were demolished when the area was redeveloped in the 1980s.

Timber, fruit, vegetables and various forms of alcohol also passed through Millwall Dock regularly, while Fred Olsen Lines combined fruit importing with passenger transport in the 1960s, taking people out to the Canary Islands and bringing back fruit and tomatoes. In the late 1980s, one of Olsen's old fruit warehouses was converted into the short-lived London Arena, which has since been replaced by residential buildings. A second Millwall fruit importer was responsible for another of the area's great legacies when, in 1885, workers from a jam factory formed Millwall FC.

The western end of the Outer Dock was originally connected to the Thames via a large entrance lock that was filled in in the 1920s, around the same time as Millwall Passage was dug to connect Millwall Dock and the West India Docks (see p.184). Nowadays, the Inner Dock is a business district while the Outer Dock is largely residential.

Chain pulleys and dock cranes

Deptford Dockyard

The dockyard at Deptford was one of the first of the Royal
Navy Dockyards, a select group of shipyards that also included
Chatham and Woolwich (see p.190).

Established in 1513 by Henry VIII as the
King's Yard, Deptford Dockyard was a chief location in Britain's
emerging shipbuilding industry. It produced vessels that fought
against the Spanish Armada in 1588 and against the French and
Spanish at the Battle of Trafalgar in 1805, as well as Captain James
Cook's favourite ship, HMS *Resolution*, in 1771. A victualling yard
beside the docks dealt in vital military supplies such as clothing,
food, tobacco and rum.

Deptford Dockyard's decline coincided with a
decline in demand for warships and ship repair after the
Napoleonic Wars at the start of the nineteenth century. The
dockyard's problems were exacerbated by its relatively shallow

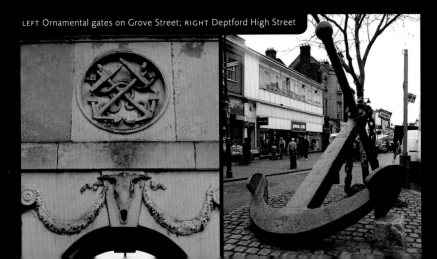

LEFT Ornamental gates on Grove Street; RIGHT Deptford High Street

A remaining dockyard slip from the Thames

water, which was entirely unsuitable for the increasingly large vessels required for international trade.

After being converted into a cattle import market between 1871 and the First World War, the dockyard was once again put to military use. During both wars, the army and its allies stored equipment there, including the amphibious vehicles that supported the D-Day landings in 1944.

Like much of London's Docklands, Deptford was heavily bombed in the early 1940s and never really recovered. The area was renamed Convoys Wharf and intermittently occupied by warehouses but it now stands derelict awaiting redevelopment.

Peter the Great at Deptford Creek

Drake's Steps

In 1967, Elizabeth II knighted record-breaking circumnavigator Francis Chichester at Greenwich in a ceremony that intentionally mirrored Drake's, including the use of the same ceremonial sword.

Deptford was home to the diarist John Evelyn in the second half of the seventeenth century, and fellow diarist Samuel Pepys often visited the docks in his official capacity as Clerk of the Acts to the Navy Board. Evelyn's house, Sayes Court (now buried beneath Convoys Wharf), achieved even greater notoriety after he sold it: on a three-month trip to learn about shipbuilding methods, Tsar Peter the Great of Russia lived at Sayes Court and trashed the place with his riotous living.

The West India Docks

The West India Docks, a group of three docks on the Isle of Dogs, were London's first purpose-built cargo-handling docks. When they opened in 1802, they constituted the largest structure of their kind in the world. A third dock – South Dock – opened in the 1860s, when the West India Dock Company converted the failed City Canal (see p.132).

The West India Dock Company was set up by Parliament at the turn of the nineteenth century after concerted lobbying by Robert Milligan, who had grown up on his family's Jamaican sugar plantations and returned to London in order to trade with the West Indies. Produce from that part of the world – including rum, coffee and sugar – was in such high demand at the time that cargo was frequently stolen upon arrival at London's sprawling docks. Milligan's plan not only ensured greater vigilance over incoming cargo but also granted him a profitable monopoly, for it decreed that all West Indian cargo into London be unloaded at the West India Docks for the next 21 years.

Records indicate that, in its five years of operation before the abolition of the slave trade, the West India Docks dispatched 77 ships to Africa in order to pick up some 25,000 slaves and transport them to the West Indies, from which the ships would return laden with goods.

The docks initially generated huge profits but these declined after the monopoly expired, and the later advent of containerisation effectively killed off the Docklands. The fortunes of the West India Docks were reversed in the 1980s and 1990s, however, with the advent of Canary Wharf and the Docklands Light Railway.

The 1807 Dockmaster's House has been used as an excise house, an office, a pub and an Indian restaurant, but was never the house of a dockmaster

Robert Milligan stands guard at an original warehouse, now the Museum of London Docklands

ROBT MILLIGAN

The East India Docks

The East India Docks were a collection of docks above the Isle of Dogs, just to the west of where the Lea joins the Thames. They consisted of an Import Dock alongside a refurbished Export Dock, which had previously existed as a ship-repairing yard called Brunswick Dock.

The East India Docks were commissioned in 1803, following the successful opening of the West India Docks (see p.184) the previous year. Their construction was heavily supported by the East India Company, which had been established in 1600 by Elizabeth I and achieved the height of its notoriety in December 1773, when its monopoly over the American tea trade provoked the Boston Tea Party.

The entrance lock on the Thames

Part of the Import Dock remains as a water feature between office blocks

The docks specialised in high-value commodities such as spices, silk, Persian carpets and of course tea. Local traders made the most of the exotic products that were coming in, setting up spice merchants and pepper grinders around the quay.

The inevitable decline that accompanied the advent of both large ships and containerisation was particularly rapid for the East India Docks because of their small size. Although the East India Dock Company had merged with the West India Dock Company in the 1830s and even built the more accessible Tilbury Docks out in Essex, trade continued to dwindle. A period of renewed industry during the Second World War was cut short by extensive bomb damage to the site, and the docks closed in 1967.

The Royal Docks

The Royal Docks were a vast docking complex north of Woolwich Reach. They comprised three interconnected docks: the Royal Victoria Dock, the Royal Albert Dock and the King George V Dock.

The first of the docks to be constructed was the Royal Victoria Dock, which opened in 1855. It was followed in 1880 by the Royal Albert Dock and in 1921 by the King George V. Plans to build a fourth dock in Beckton never came to fruition but the Royal Docks were nevertheless the largest enclosed docks in the world, with a combined capacity of 250 acres of water.

In the first half of the twentieth century – a time when many of London's smaller docks were facing financial disaster and abandonment – the Royal Docks thrived. Supported by rows of sheds and

Remains of dock buildings and crane tracks at the King George V Dock

London City Airport's runway between the King George V Dock and the Royal Albert Dock

Woolwich Dockyard

Woolwich Dockyard was one of the earliest Royal Navy Dockyards established in London, forming a pair with Deptford Dockyard (see p.180) on either side of Henry VIII's beloved Greenwich.

The first great achievement at Woolwich Dockyard was the launch in 1514 of the *Henry Grace à Dieu*, nicknamed the *Great Harry*, a large and very mighty-looking warship that ultimately saw more service as a ceremonial vehicle for Henry VIII than as a military vessel. Its lasting legacy to British seafaring, however, is the HMS *Beagle*, launched from Woolwich in 1820, in which Charles Darwin travelled the world in the early 1830s collecting species in support of his emerging theory of evolution.

The dockyard's main reputation was for repair work, in which it was supported from the 1570s onwards by an enormous ropeyard, but it also launched well over a dozen high-profile royal vessels during its years of operation. Its work was valuable enough to warrant frequent visits from Samuel Pepys, Clerk of the Acts to the Navy Board, during the 1660s, as well as defences including cannons positioned around its perimeter.

Woolwich Dockyard fell into decline in the nineteenth century when it could no longer cope with the ever-increasing size of ships, while the silting-up of the Thames made it almost impossible for the dockyard to accept commissions. It closed down in 1869, 20 years after giving its name to the local railway station. Dockyard buildings now sit among council properties and the remaining strips of water are used for angling.

Defences at the Thames entrance

The former dock is now a fishing lake

Pudding Mill River